Fundamental Theories of Physics

Volume 173

For further volumes:
www.springer.com/series/6001

Ruben Aldrovandi • José Geraldo Pereira

Teleparallel Gravity

An Introduction

Springer

Ruben Aldrovandi
Instituto de Física Teórica
Universidade Estadual Paulista
São Paulo, Brazil

José Geraldo Pereira
Instituto de Física Teórica
Universidade Estadual Paulista
São Paulo, Brazil

ISBN 978-94-007-5142-2 ISBN 978-94-007-5143-9 (eBook)
DOI 10.1007/978-94-007-5143-9
Springer Dordrecht Heidelberg New York London

Library of Congress Control Number: 2012945747

Printed on acid-free paper

Springer is part of Springer Science+Business Media (www.springer.com)

To our parents

Preface

Gravitation stands apart among the four fundamental interactions of Nature. Its source is energy-momentum, a current every field or body gives rise to. It stands particularly distinctive in its description by General Relativity, in which the gravitational effects remain amalgamated with inertia. Furthermore, from the theoretical point of view, General Relativity is the only fundamental interaction theory which does not adhere to the gauge template. Teleparallel Gravity, the theory this book is devoted to, is an alternative description that, though equivalent to General Relativity, separates gravitation from inertia and meets the requirements of a gauge theory.

The simplest example of a gauge theory is electromagnetism, which is, for this reason, quite appropriate for understanding the basic tenets of the gauge paradigm. A fundamental piece of this paradigm is Noether's theorem. According to it, the electric current, the source of the electromagnetic field, is conserved provided the action functional of the source field be invariant under the transformations of the unitary group $U(1)$, the gauge group of the theory. In the same token, the source of the gravitational field is well known to be energy and momentum. Noether's theorem will say that the energy-momentum current tensor is conserved provided the source action functional be invariant under spacetime translations. If gravitation is to present a gauge formulation with energy-momentum as source, it must then be a gauge theory for the translation group. This is precisely Teleparallel Gravity.

Although equivalent to General Relativity, Teleparallel Gravity is, conceptually speaking, a completely different theory. In General Relativity, curvature is used to geometrize the gravitational interaction: geometry replaces the concept of gravitational force, and the trajectories are determined, not by force equations, but by geodesics. Teleparallel Gravity, on the other hand, attributes gravitation to torsion, but not through a geometrization: it acts through a force. In consequence, there are no geodesic equations in it, only force equations quite analogous to the Lorentz force equation of electrodynamics.

The reason for gravitation to present two equivalent descriptions lies in its most peculiar property: universality. Like the other fundamental interactions of Nature, gravitation can be described in terms of a gauge theory, which is just Teleparallel Gravity. Universality of free fall, on the other hand, makes it possible a second, geometrized description, based on the equivalence principle, which is just General Relativity. As the sole universal interaction, it is the only one to allow also a geometrical interpretation—hence the possibility of two descriptions. From this point of view, curvature and torsion are simply alternative ways of representing the same gravitational field, accounting for the same gravitational degrees of freedom.

The teleparallel structure was already known in the nineteen-twenties, and was used by Einstein in his unsuccessful attempt to unify electromagnetism and gravitation. The birth of teleparallelism as a gravitational theory, however, took place in the nineteen-fifties with the works by Møller. Since then, there have been many contributions from different authors to Teleparallel Gravity [see Sect. 4.1 for a brief historical review]. Although its foundations can be considered quite well understood by now, there are still some different interpretations for the role played by torsion in the description of the gravitational interaction. It is thus important to keep in mind that the ideas presented here are strongly biased by the authors' point of view on the subject, and are essentially based on the research developed by them along many years. In this sense this book is partially an expanded version of their relevant publications, and includes also many corrections to the original texts. Even though the authors are the solely responsible for the contents of the book, they owe much to some of their colleagues, as well as to their former and present collaborators: V.C. de Andrade, H.I. Arcos, T.V. Aucalla, A.L. Barbosa, P.B. Barros, T. Gribl Lucas, F.W. Hehl, L.C.T. Guillen, J.W. Maluf, R.A. Mosna, J. Nester, Yu.N. Obukhov, D.J. Rezende, R. da Rocha, G. Rubilar, K.H. Vu, P. Zambianchi and C.M. Zhang. The authors would like to express their deep gratitude to all of them. Finally, they wish to thank Aldo Rampioni, publishing editor of Theoretical & Mathematical Physics and Scientific Computation, Vesselin Petkov, member of the editorial board of the series *Fundamental Theories of Physics*, and Kirsten Theunissen from Springer for their friendly assistance and support during the completion of the book.

São Paulo, Brazil

Ruben Aldrovandi
José Geraldo Pereira

Contents

Chapter 1
Basic Notions

A general spacetime is a 4-dimensional differentiable manifold whose tangent space is, at each point, a Minkowski spacetime. Linear frames and tetrad fields are constitutive parts of its structure as a manifold, and instrumental in relativistic physics and gravitation. They are defined up to point-dependent Lorentz transformations, under which usual derivatives exhibit a non-covariance that can be just compensated by the non-covariance of connections, objects thereby essential to produce meaningful, covariant derivatives. Each connection defines a covariant derivative, from which two basic covariant objects result: curvature and torsion. These quantities satisfy two mandatory relations, the Bianchi identities.

1.1 Linear Frames and Tetrads

Spacetime is the common arena on which the four presently known fundamental interactions manifest themselves. Electromagnetic, weak and strong interactions are currently described by gauge theories involving transformations taking place in *internal* spaces, by themselves unrelated to spacetime. Gravitation, on the other hand, is deeply linked with the very structure of spacetime.

The theories describing the four interactions have all a strong geometrical flavor. For gauge theories, the basic settings are principal bundles[1] with a copy of the corresponding gauge group at each spacetime point. The geometrical setting of any theory for gravitation is the tangent bundle, a natural construction always present in any differentiable manifold [1]. In fact, at each point of spacetime—the base space of the bundle—there is always a tangent space attached to it—the fiber of the bundle, which is seen as a vector space.

Comment 1.1 Mathematicians use an *invariant* language, stating and proving results without any use of explicit bases or coordinates. Physicists use a *covariant* language, in part because they are used to, but mainly because they have to prepare for experiments, which are always performed in a particular frame, using apparatuses which suppose a particular coordinate system. Also, physicists have a more pictorial discourse, frequently frowned upon by mathematicians, in which immediate

[1] Bundles will be discussed in some more detail in Chap. 3.

R. Aldrovandi, J.G. Pereira, *Teleparallel Gravity*, Fundamental Theories of Physics 173,
DOI 10.1007/978-94-007-5143-9_1,

intuition plays a dominant role. For example, tangent spaces are spoken of as "touching" a manifold at a point, internal spaces are "attached" to the manifold at a point, etc. We shall, of course, follow this practice.

We are going to use the Greek alphabet ($\mu, \nu, \rho, \ldots = 0, 1, 2, 3$) to denote indices related to spacetime, and the first letters of the Latin alphabet ($a, b, c, \ldots = 0, 1, 2, 3$) to denote indices related to the tangent space, a Minkowski spacetime whose Lorentz metric is assumed to have the form

$$\eta_{ab} = \mathrm{diag}(+1, -1, -1, -1) \tag{1.1}$$

in cartesian coordinates. The middle letters of the Latin alphabet ($i, j, k, \ldots = 1, 2, 3$) will be reserved for space indices.

A general spacetime is a 4-dimensional differential manifold (indicated $\mathbb{R}^{3,1}$ from now on) whose tangent space is, at each point, a Minkowski spacetime. The motions on each one of these tangent spaces are the transformations constituting the Poincaré group

$$\mathscr{P} = \mathscr{L} \oslash \mathscr{T}^{3,1}, \tag{1.2}$$

the semi-direct product of the Lorentz $\mathscr{L} = SO(3, 1)$ group (including rotations and boosts) by the group of translations $\mathscr{T}^{3,1}$.

Comment 1.2 It is important to remark that this does not necessarily mean that the gauge group of gravitation is the Poincaré group. See Comment 5.1 for a further discussion of this point.

Spacetime coordinates will be denoted by $\{x^\mu\}$, whereas tangent space coordinates will be denoted by $\{x^a\}$. Such coordinate systems determine, on their domains of definition, local bases for vector fields, formed by the sets of gradients

$$\{\partial_\mu\} \equiv \left\{\partial/\partial x^\mu\right\} \quad \text{and} \quad \{\partial_a\} \equiv \left\{\partial/\partial x^a\right\}, \tag{1.3}$$

as well as bases $\{dx^\mu\}$ and $\{dx^a\}$ for covector fields, or differentials. These bases are dual, in the sense that

$$dx^\mu(\partial_\nu) = \delta^\mu_\nu \quad \text{and} \quad dx^a(\partial_b) = \delta^a_b. \tag{1.4}$$

On the respective domains of definition, any vector or covector can be expressed in terms of these bases, which can furthermore be extended by direct product to constitute bases for general tensor fields of any order.

1.1.1 Trivial Frames

General frames, or tetrads—also frequently called *vierbeine*, or four-legs—will be denoted by

$$\{e_a\} \quad \text{and} \quad \left\{e^a\right\}. \tag{1.5}$$

Very particular cases are the mentioned "coordinate" bases

$$\{e_a\} = \{\partial_a\} \quad \text{and} \quad \{e^a\} = \{dx^a\}, \tag{1.6}$$

whose name stems from their relationship to a coordinate system. Any other set of four linearly independent fields $\{e_a\}$ will form another basis, and will have a dual $\{e^a\}$ whose members are such that

$$e^a(e_b) = \delta^a_b. \tag{1.7}$$

Notice that, on a general manifold, vector fields are (like coordinate systems) only locally defined—and linear frames, as sets of four such fields, are only defined on restricted domains.

Comment 1.3 The rather special manifolds on which a vector field can be defined everywhere are called *parallelizable*. Euclidean spaces $\mathbb{E}^n$ are the simplest examples. Of all the spheres $\mathbb{S}^n$, only $\mathbb{S}^1$, $\mathbb{S}^3$ and $\mathbb{S}^7$ are parallelizable [2]. Lie groups are parallelizable manifolds, which means, for instance, that no Lie group can have $\mathbb{S}^2$ for its underlying manifold. Also all toruses are parallelizable. A vector field on a non-parallelizable manifold will always vanish at some point, which is quite unacceptable for the member of a vector basis.

These frame fields are the general linear bases on the spacetime differentiable manifold $\mathbb{R}^{3,1}$. The whole set of such bases, under conditions making of it also a differentiable manifold, constitutes the *bundle of linear frames*. A frame field provides, at each point $p \in \mathbb{R}^{3,1}$, a basis for the vectors on the tangent space $T_p\mathbb{R}^{3,1}$. Of course, on the common domains they are defined, each member of a given basis can be written in terms of the members of any other. For example,

$$e_a = e_a{}^{\mu}\partial_{\mu} \quad \text{and} \quad e^a = e^a{}_{\mu}\, dx^{\mu}, \tag{1.8}$$

and conversely,

$$\partial_{\mu} = e^a{}_{\mu} e_a \quad \text{and} \quad dx^{\mu} = e_a{}^{\mu} e^a. \tag{1.9}$$

On account of the orthogonality conditions (1.7), the frame components satisfy

$$e^a{}_{\mu} e_a{}^{\nu} = \delta^{\nu}_{\mu} \quad \text{and} \quad e^a{}_{\mu} e_b{}^{\mu} = \delta^a_b. \tag{1.10}$$

Notice that these frames, with their bundles, are constitutive parts of spacetime: they are automatically present as soon as spacetime is taken to be a differentiable manifold.

A general linear basis $\{e_a\}$ satisfies the commutation relation

$$[e_a, e_b] = f^c{}_{ab} e_c, \tag{1.11}$$

with $f^c{}_{ab}$ the so-called structure coefficients, or coefficients of anholonomy, or still the anholonomy of frame $\{e_a\}$. The dual expression of the commutation relation above is the Cartan structure equation

$$de^c = -\tfrac{1}{2} f^c{}_{ab} e^a \wedge e^b = \tfrac{1}{2}\left(\partial_{\mu} e^c{}_{\nu} - \partial_{\nu} e^c{}_{\mu}\right) dx^{\mu} \wedge dx^{\nu}. \tag{1.12}$$

The structure coefficients represent the curls of the basis members:

$$f^c{}_{ab} = e^c{}_{\mu}\left[e_a\left(e_b{}^{\mu}\right) - e_b\left(e_a{}^{\mu}\right)\right] = e_a{}^{\mu} e_b{}^{\nu}\left(\partial_{\nu} e^c{}_{\mu} - \partial_{\mu} e^c{}_{\nu}\right). \tag{1.13}$$

A preferred class is that of inertial frames, denoted e'_a, those for which

$$f'^{a}{}_{cd} = 0. \tag{1.14}$$

Notice that $f'^{c}{}_{ab} = 0$ means $de'^{a} = 0$, which in turn implies that e'^{a} is a closed differential form, consequently locally exact: $e'^{a} = dx^a$ for some x^a. Basis $\{e'^{a}\}$ is then said to be integrable, or *holonomic*. Of course, all coordinate bases are holonomic. This is not a local property, in the sense that it is valid everywhere for frames belonging to this inertial class.

Comment 1.4 Anholonomy—the property by which a differential form is not the differential of anything, or of a vector field which is not a gradient—is commonplace in many chapters of Physics. Heat and work, for instance, are typical anholonomic coordinates on the space of thermodynamic variables, and the angular velocity of a generic rigid body in euclidean space $\mathbb{E}^3$ is a classical example of anholonomic velocity.

Consider now the Minkowski spacetime metric which, written in a holonomic basis $\{dx^\mu\}$, reads

$$\boldsymbol{\eta} = \eta_{\mu\nu}\, dx^\mu \otimes dx^\nu = \eta_{\mu\nu}\, dx^\mu\, dx^\nu. \tag{1.15}$$

When $\{x^\mu\}$ represents a set of cartesian coordinates, it has the form

$$\eta_{\mu\nu} = \text{diag}(+1, -1, -1, -1). \tag{1.16}$$

In any other coordinates, $\eta_{\mu\nu}$ will be a function of the spacetime coordinates. The linear frame

$$e_a = e_a{}^\mu \partial_\mu, \tag{1.17}$$

provides a relation between the tangent-space metric

$$\boldsymbol{\eta} = \eta_{ab}\, dx^a \otimes dx^b = \eta_{ab}\, dx^a dx^b, \tag{1.18}$$

and the spacetime metric $\eta_{\mu\nu}$. Such relation is given by

$$\eta_{ab} = \eta_{\mu\nu} e_a{}^\mu e_b{}^\nu. \tag{1.19}$$

Using the orthogonality conditions (1.10), the inverse relation is found to be

$$\eta_{\mu\nu} = \eta_{ab} e^a{}_\mu e^b{}_\nu. \tag{1.20}$$

Independently of whether e_a is holonomic or not or, equivalently, whether they are inertial or not, they always relate the tangent Minkowski space to a Minkowski spacetime. These are the frames appearing in Special Relativity, and are usually called trivial frames, or trivial tetrads.

Comment 1.5 A *caveat*: there is no consensus on the topology of spacetime. This means that frequently used notions like "neighborhood", "coordinate" and "continuity" are actually not well defined from the mathematical point of view. The Lorentz metric, being non-positive definite, does not define any topology [3]: its role is actually to introduce causality. Many proposals have been made to fix that topology [4–10], but none has obtained general acceptance. In practice, physicists make implicitly a purely operational option: they use an underlying euclidean $\mathbb{E}^4$ when eventually using global coordinates, or when talking about "continuous" fields. In order to have causality, they then superpose an *additional* Lorentz metric, making of $\mathbb{E}^4$ a Minkowski $\mathbb{M} \equiv \mathbb{E}^{3,1}$. This is the causal space. They finally deform $\mathbb{E}^{3,1}$ into a pseudo-riemannian space $\mathbb{R}^{3,1}$ of the same signature, so as to locally preserve causality. This $\mathbb{R}^{3,1}$ has, at each point, a tangent space which is identical to the causal Minkowski $\mathbb{M}$.

1.1.2 Nontrivial Frames

Nontrivial frames (or tetrads) will be denoted by

$$\{h_a\} \quad \text{and} \quad \{h^a\}. \tag{1.21}$$

They are defined as linear frames whose coefficient of anholonomy is related to both inertia *and* gravitation. To see the difference with relation to the trivial, non-gravitational linear frames e_a, let us consider a general pseudo-riemannian space-time metric $\boldsymbol{g}$, with components $g_{\mu\nu}$ in the some dual holonomic basis $\{dx^\mu\}$:

$$\boldsymbol{g} = g_{\mu\nu}\, dx^\mu \otimes dx^\nu = g_{\mu\nu}\, dx^\mu\, dx^\nu. \tag{1.22}$$

The tetrad field

$$h_a = h_a{}^\mu \partial_\mu \quad \text{and} \quad h^a = h^a{}_\mu dx^\mu \tag{1.23}$$

is a linear basis that relates $\boldsymbol{g}$ to the tangent-space metric

$$\boldsymbol{\eta} = \eta_{ab}\, dx^a \otimes dx^b = \eta_{ab}\, dx^a\, dx^b \tag{1.24}$$

through the relation

$$\eta_{ab} = g_{\mu\nu} h_a{}^\mu h_b{}^\nu. \tag{1.25}$$

By (1.1), this means that a tetrad field is a linear frame whose members h_a are pseudo-orthogonal by the pseudo-riemannian metric $g_{\mu\nu}$. The components of the dual basis members $h^a = h^a{}_\nu dx^\nu$ satisfy

$$h^a{}_\mu h_a{}^\nu = \delta^\nu_\mu \quad \text{and} \quad h^a{}_\mu h_b{}^\mu = \delta^a_b, \tag{1.26}$$

so that Eq. (1.25) has the converse

$$g_{\mu\nu} = \eta_{ab} h^a{}_\mu h^b{}_\nu. \tag{1.27}$$

Notice that the Lorentz metric (1.1) fixes the signature s (the number of positive eigenvalues minus the number of negative eigenvalues) as $s = 2$ for all metrics given by (1.27). The determinant

$$g = \det(g_{\mu\nu}) \tag{1.28}$$

is negative because of the signature of metric η_{ab}. We shall also be using the notation

$$h = \det\left(h^a{}_\mu\right) = \sqrt{-g}. \tag{1.29}$$

A tetrad basis $\{h_a\}$ satisfies the commutation relation

$$[h_a, h_b] = f^c{}_{ab} h_c, \tag{1.30}$$

with $f^c{}_{ab}$ the structure coefficients, or coefficients of anholonomy, of frame $\{h_a\}$. The basic difference in relation to the linear bases $\{e_a\}$ is that now the $f^c{}_{ab}$ represent both inertia and gravitation. The dual expression of the commutation relation above is the Cartan structure equation, now for $\{h_a\}$:

$$dh^c = -\tfrac{1}{2} f^c{}_{ab}\, h^a \wedge h^b = \tfrac{1}{2}\left(\partial_\mu h^c{}_\nu - \partial_\nu h^c{}_\mu\right) dx^\mu \wedge dx^\nu. \tag{1.31}$$

As before, the structure coefficients are given by the curls of the basis members:

$$f^c{}_{ab} = h^c{}_\mu\big[h_a\big(h_b{}^\mu\big) - h_b\big(h_a{}^\mu\big)\big] = h_a{}^\mu h_b{}^\nu\big(\partial_\nu h^c{}_\mu - \partial_\mu h^c{}_\nu\big). \tag{1.32}$$

Although tetrads are, by definition, anholonomic due to the presence of gravitation, it is still possible that *locally*, $f^c{}_{ab} = 0$. In this case, $dh^a = 0$, which means that h^a is locally a closed differential form. In fact, if this holds at a point p, then there is a neighborhood around p on which functions (coordinates) x^a exist such that

$$h^a = dx^a.$$

We say that a closed differential form is always locally integrable, or exact. This is the case of inertial frames, which are always holonomic.

Comment 1.6 Differently from the special relativistic case, where $f^c{}_{ab} = 0$ means absence of inertial effects, in the presence of gravitation, since $f^c{}_{ab}$ represents both inertial effects and gravitation, the local vanishing $f^c{}_{ab} = 0$ means a local frame in which inertial effects exactly compensate gravitation.

1.2 Lorentz Connections

Objects with a well-defined behavior under point-dependent transformations (general coordinate and gauge transformations, for example) are rather loosely called *covariant* under those transformations. Ordinary derivatives of such covariant objects are not themselves covariant. In order to define derivatives with a well-defined tensor behavior (that is, which are covariant), it is essential to introduce connections A_μ, which behave like vectors in what concerns the spacetime index, but whose non-tensorial behavior in the algebraic indices just compensates the non-tensoriality of ordinary derivatives. Gauge potentials (for example, the electromagnetic potential) are connections, introduced to produce derivatives that are covariant under gauge transformations—that is, point-dependent transformations taking place in *internal* spaces (of isospin, of color, etc.).

Connections related to the linear group $GL(4, \mathbb{R})$ and its subgroups—such as the Lorentz group $SO(3, 1)$—are called *linear* connections. They have a larger degree of intimacy with spacetime because they are defined on the bundle of linear frames, which is a constitutive part of its manifold structure. That bundle has some properties not found in the bundles related to *internal* gauge theories. Mainly, it exhibits soldering, which leads to the existence of *torsion* for every connection. Linear connections—in particular, Lorentz connections—always have torsion, while internal gauge potentials do not have.

Comment 1.7 It is worth noticing that a vanishing torsion is quite different from a non-existent (non-defined) torsion. In fact, we shall see below that the zero torsion that appears in General Relativity has an important consequence: the cyclic symmetry of the curvature tensor [see Eq. (1.88)], for which no analog exists in internal gauge theories.

A *Lorentz connection* A_μ, frequently referred to also as *spin connection*, is a 1-form assuming values in the Lie algebra of the Lorentz group,

$$A_\mu = \tfrac{1}{2} A^{ab}{}_\mu S_{ab}, \tag{1.33}$$

with S_{ab} a given representation of the Lorentz generators. As these generators are antisymmetric in the algebraic indices, $A^{ab}{}_\mu$ must be equally antisymmetric in order to be lorentzian. This connection defines the Fock-Ivanenko covariant derivative [11, 12]

$$\mathscr{D}_\mu = \partial_\mu - \tfrac{i}{2} A^{ab}{}_\mu S_{ab}, \tag{1.34}$$

whose second part acts only on the algebraic, or tangent space indices. For a scalar field ϕ, for example, the generators are [13]

$$S_{ab} = 0. \tag{1.35}$$

For a Dirac spinor ψ, they are spinorial matrices of the form

$$S_{ab} = \tfrac{i}{4} [\gamma_a, \gamma_b], \tag{1.36}$$

with γ_a the Dirac matrices. A Lorentz vector field ϕ^c, on the other hand, is acted upon by the vector representation of the Lorentz generators, matrices S_{ab} with entries

$$(S_{ab})^c{}_d = i\left(\eta_{bd}\delta^c_a - \eta_{ad}\delta^c_b\right). \tag{1.37}$$

The Fock-Ivanenko derivative is, in this case,

$$\mathscr{D}_\mu \phi^c = \partial_\mu \phi^c + A^c{}_{d\mu} \phi^d, \tag{1.38}$$

and so on for any other fundamental field.

On account of the soldered character of the tangent bundle, a tetrad field relates tangent space (or internal) tensors with spacetime (or external) tensors. For example, if ϕ^a is an internal, or Lorentz vector, then

$$\phi^\rho = h_a{}^\rho \phi^a \tag{1.39}$$

will be a spacetime vector. Conversely, we can write

$$\phi^a = h^a{}_\rho \phi^\rho. \tag{1.40}$$

On the other hand, due to its non-tensorial character, a connection will acquire a vacuum, or non-homogeneous term, under the same operation. For example, to each spin connection $A^a{}_{b\mu}$, there is a corresponding general linear connection $\Gamma^\rho{}_{\nu\mu}$, given by

$$\Gamma^\rho{}_{\nu\mu} = h_a{}^\rho \partial_\mu h^a{}_\nu + h_a{}^\rho A^a{}_{b\mu} h^b{}_\nu \equiv h_a{}^\rho \mathscr{D}_\mu h^a{}_\nu, \tag{1.41}$$

where $\mathscr{D}_\mu$ is the covariant derivative (1.38), in which the generators act on internal (or tangent space) indices only. The inverse relation is, consequently,

$$A^a{}_{b\mu} = h^a{}_\nu \partial_\mu h_b{}^\nu + h^a{}_\nu \Gamma^\nu{}_{\rho\mu} h_b{}^\rho \equiv h^a{}_\nu \nabla_\mu h_b{}^\nu, \tag{1.42}$$

where ∇_μ is the standard covariant derivative in the connection $\Gamma^\nu{}_{\rho\mu}$, which acts on external indices only. For a spacetime vector ϕ^ν, for example, it is given by

$$\nabla_\mu \phi^\nu = \partial_\mu \phi^\nu + \Gamma^\nu{}_{\rho\mu} \phi^\rho. \tag{1.43}$$

Using relations (1.39) and (1.40), it is easy to verify that

$$\mathscr{D}_\mu \phi^d = h^d{}_\rho \nabla_\mu \phi^\rho. \tag{1.44}$$

Comment 1.8 Observe that, whereas the Fock-Ivanenko derivative $\mathscr{D}_\mu$ can be defined for all fields—tensorial and spinorial—the covariant derivative ∇_μ can be defined for tensorial fields only. In order to describe the interaction of spinor fields with gravitation, therefore, the use of Fock-Ivanenko derivatives is mandatory [14, 15].

Equations (1.41) and (1.42) are simply different ways of expressing the property that the total covariant derivative of the tetrad—that is, a covariant derivative with connection terms for both internal and external indices—vanishes identically:

$$\partial_\mu h^a{}_\nu - \Gamma^\rho{}_{\nu\mu} h^a{}_\rho + A^a{}_{b\mu} h^b{}_\nu = 0. \tag{1.45}$$

On the other hand, a connection $\Gamma^\rho{}_{\lambda\mu}$ is said to be metric compatible if the so-called *metricity condition*

$$\nabla_\lambda g_{\mu\nu} \equiv \partial_\lambda g_{\mu\nu} - \Gamma^\rho{}_{\mu\lambda} g_{\rho\nu} - \Gamma^\rho{}_{\nu\lambda} g_{\mu\rho} = 0 \tag{1.46}$$

is fulfilled. From the tetrad point of view, and using Eqs. (1.41) and (1.42), this equation can be rewritten in the form

$$\partial_\mu \eta_{ab} - A^d{}_{a\mu} \eta_{db} - A^d{}_{b\mu} \eta_{ad} = 0, \tag{1.47}$$

or equivalently

$$A_{ba\mu} = -A_{ab\mu}. \tag{1.48}$$

The underlying content of the metric-preserving property, therefore, is that the spin connection is lorentzian—that is, antisymmetric in the algebraic indices. Conversely, when $\nabla_\lambda g_{\mu\nu} \neq 0$, the corresponding connection $A^a{}_{b\mu}$ does not assume values in the Lie algebra of the Lorentz group—it is not a Lorentz connection.

1.3 Curvature and Torsion

Curvature and torsion are tensorial properties of Lorentz connections [1]. This becomes evident if we note that many different connections can be defined on the very same metric spacetime [2]. Of course, when restricted to the specific case of General Relativity, where only the zero-torsion spin connection is present, universality of gravitation allows its curvature to be interpreted—together with the metric—as part of the spacetime definition, and one can then talk about "spacetime curvature". However, in the presence of connections with different curvatures and torsions, it seems far more convenient to follow the mathematicians and take spacetime simply as a manifold, and connections (with their curvatures and torsions) as additional structures.

The curvature of a Lorentz connection $A^a{}_{b\mu}$ is a 2-form assuming values in the Lie algebra of the Lorentz group,

$$R = \tfrac{1}{4} R^a{}_{b\nu\mu} S_a{}^b \, dx^\nu \wedge dx^\mu. \tag{1.49}$$

Torsion is also a 2-form, but assuming values in the Lie algebra of the translation group,

$$T = \tfrac{1}{2} T^a{}_{\nu\mu} P_a \, dx^\nu \wedge dx^\mu, \tag{1.50}$$

with $P_a = \partial_a$ the translation generators. The curvature and torsion components are defined, respectively, by

$$R^a{}_{b\nu\mu} = \partial_\nu A^a{}_{b\mu} - \partial_\mu A^a{}_{b\nu} + A^a{}_{e\nu} A^e{}_{b\mu} - A^a{}_{e\mu} A^e{}_{b\nu} \tag{1.51}$$

and

$$T^a{}_{\nu\mu} = \partial_\nu h^a{}_\mu - \partial_\mu h^a{}_\nu + A^a{}_{e\nu} h^e{}_\mu - A^a{}_{e\mu} h^e{}_\nu. \tag{1.52}$$

Through contraction with tetrads, these tensors can be written in spacetime-indexed forms:

$$R^\rho{}_{\lambda\nu\mu} = h_a{}^\rho h^b{}_\lambda R^a{}_{b\nu\mu}, \tag{1.53}$$

and

$$T^\rho{}_{\nu\mu} = h_a{}^\rho T^a{}_{\nu\mu}. \tag{1.54}$$

Using relation (1.42), their components are found to be

$$R^\rho{}_{\lambda\nu\mu} = \partial_\nu \Gamma^\rho{}_{\lambda\mu} - \partial_\mu \Gamma^\rho{}_{\lambda\nu} + \Gamma^\rho{}_{\eta\nu} \Gamma^\eta{}_{\lambda\mu} - \Gamma^\rho{}_{\eta\mu} \Gamma^\eta{}_{\lambda\nu} \tag{1.55}$$

and

$$T^\rho{}_{\nu\mu} = \Gamma^\rho{}_{\mu\nu} - \Gamma^\rho{}_{\nu\mu}. \tag{1.56}$$

Considering that the spin connection $A^a{}_{b\nu}$ is a (co-)vector in the last index, we can write

$$A^a{}_{bc} = A^a{}_{b\nu} h_c{}^\nu. \tag{1.57}$$

It can thus be verified that, in the anholonomic basis $\{h_a\}$, the curvature and torsion components are given respectively by

$$R^a{}_{bcd} = h_c\big(A^a{}_{bd}\big) - h_d\big(A^a{}_{bc}\big) + A^a{}_{ec} A^e{}_{bd} - A^a{}_{ed} A^e{}_{bc} - f^e{}_{cd} A^a{}_{be} \tag{1.58}$$

and

$$T^a{}_{bc} = A^a{}_{cb} - A^a{}_{bc} - f^a{}_{bc}, \tag{1.59}$$

where, we recall, $h_c = h_c{}^\mu \partial_\mu$. Use of (1.59) for three different combinations of indices gives

$$A^a{}_{bc} = \tfrac{1}{2}\big(f_b{}^a{}_c + T_b{}^a{}_c + f_c{}^a{}_b + T_c{}^a{}_b - f^a{}_{bc} - T^a{}_{bc}\big). \tag{1.60}$$

This expression can be rewritten in the form[2]

$$A^a{}_{bc} = \overset{\circ}{A}{}^a{}_{bc} + K^a{}_{bc}, \tag{1.61}$$

where

$$\overset{\circ}{A}{}^a{}_{bc} = \tfrac{1}{2}\big(f_b{}^a{}_c + f_c{}^a{}_b - f^a{}_{bc}\big) \tag{1.62}$$

is the usual expression of the General Relativity spin connection in terms of the coefficients of anholonomy, and

$$K^a{}_{bc} = \tfrac{1}{2}\big(T_b{}^a{}_c + T_c{}^a{}_b - T^a{}_{bc}\big) \tag{1.63}$$

is the *contortion tensor*. As a matter of fact, like the Lorentz connection, contortion is also a 1-form assuming values in the Lie algebra of the Lorentz group:

$$K_\mu = \tfrac{1}{2} K^a{}_{b\mu} S_a{}^b. \tag{1.64}$$

Equation (1.61) is actually the content of a theorem, which states that any Lorentz connection can be decomposed into the spin connection of General Relativity plus the contortion tensor [1]. It is a fundamental result that will be used many times along this book. In terms of the corresponding spacetime-indexed linear connection, it reads

$$\Gamma^\rho{}_{\mu\nu} = \overset{\circ}{\Gamma}{}^\rho{}_{\mu\nu} + K^\rho{}_{\mu\nu}, \tag{1.65}$$

where

$$\overset{\circ}{\Gamma}{}^\sigma{}_{\mu\nu} = \tfrac{1}{2} g^{\sigma\rho}(\partial_\mu g_{\rho\nu} + \partial_\nu g_{\rho\mu} - \partial_\rho g_{\mu\nu}) \tag{1.66}$$

is the zero-torsion Christoffel, or Levi-Civita connection, and

$$K^\rho{}_{\mu\nu} = \tfrac{1}{2}\big(T_\nu{}^\rho{}_\mu + T_\mu{}^\rho{}_\nu - T^\rho{}_{\mu\nu}\big) \tag{1.67}$$

is the spacetime-indexed contortion tensor.

Comment 1.9 Any tensor can be decomposed into its antisymmetric and symmetric parts. The antisymmetric part is defined by

$$C_{[ab]} = \frac{1}{2!}(C_{ab} - C_{ba})$$
$$D_{[abc]} = \frac{1}{3!}(D_{abc} - D_{acb} + D_{cab} - D_{bac} + D_{bca} - D_{cba}),$$

and so on. In the same way, the symmetric part is

$$C_{(ab)} = \frac{1}{2!}(C_{ab} + C_{ba})$$
$$D_{(abc)} = \frac{1}{3!}(D_{abc} + D_{acb} + D_{cab} + D_{bac} + D_{bca} + D_{cba}),$$

and so on. For example, Eq. (1.60) can be rewritten

$$A^a{}_{bc} = A^a{}_{(bc)} + A^a{}_{[bc]}, \tag{1.68}$$

[2] All quantities related to General Relativity will be denoted with an over "∘".

with

$$A^a{}_{(bc)} = -\tfrac{1}{2}\left(f_{bc}{}^a + f_{cb}{}^a + T_{bc}{}^a + T_{cb}{}^a\right) \tag{1.69}$$

and

$$A^a{}_{[bc]} = -\tfrac{1}{2}\left(f^a{}_{bc} + T^a{}_{bc}\right). \tag{1.70}$$

From this expression we see that, given a tetrad, the connection is completely determined by its torsion—which is Ricci's theorem [16].

1.4 Local Lorentz Transformations

A local Lorentz transformation is fundamentally a transformation of the tangent space coordinates x^a:

$$x'^a = \Lambda^a{}_b(x) x^b. \tag{1.71}$$

Under such a transformation, the tetrad frames transform according to

$$h'^a = \Lambda^a{}_b(x) h^b \quad \text{and} \quad h'_a = \Lambda_a{}^b(x) h_b. \tag{1.72}$$

At each point of a riemannian spacetime, Eq. (1.27) only determines the tetrad field up to transformations of the six-parameter Lorentz group in the tangent space indices. This means that there exists actually a six-fold infinity of tetrads $h_a{}^\mu$, each one relating the spacetime metric $\boldsymbol{g}$ to the tangent space metric $\boldsymbol{\eta}$ by Eqs. (1.25) and (1.27). Suppose in effect another tetrad $\{h'_a\}$ such that

$$g_{\mu\nu} = \eta_{cd} h'^c{}_\mu h'^d{}_\nu, \tag{1.73}$$

Contracting both sides with $h_a{}^\mu h_b{}^\nu$, we arrive at

$$\eta_{ab} = \eta_{cd}\left(h'^c{}_\mu h_a{}^\mu\right)\left(h'^d{}_\nu h_b{}^\nu\right). \tag{1.74}$$

This equation says that the matrix with entries

$$\Lambda^a{}_b(x) = h'^a{}_\mu h_b{}^\mu, \tag{1.75}$$

which gives the transformation

$$h'^a{}_\mu = \Lambda^a{}_b(x) h^b{}_\mu, \tag{1.76}$$

satisfies

$$\eta_{cd} \Lambda^c{}_a(x) \Lambda^d{}_b(x) = \eta_{ab}. \tag{1.77}$$

This is just the condition that a matrix Λ must satisfy in order to belong to (the vector representation of) the Lorentz group. The converse reasoning will say that Lorentz transformations preserve the metric defined by $\{h_a\}$.

Under a local Lorentz transformation $\Lambda^a{}_b(x)$, the spin connection undergoes the transformation

$$A'^a{}_{b\mu} = \Lambda^a{}_c(x) A^c{}_{d\mu} \Lambda_b{}^d(x) + \Lambda^a{}_c(x) \partial_\mu \Lambda_b{}^c(x). \tag{1.78}$$

In the same way, it is easy to verify that $T^a{}_{\nu\mu}$ and $R^a{}_{b\nu\mu}$ transform covariantly:

$$T'^a{}_{\nu\mu} = \Lambda^a{}_b(x) T^b{}_{\nu\mu} \tag{1.79}$$

and

$$R'^a{}_{b\nu\mu} = \Lambda^a{}_c(x) \Lambda_b{}^d(x) R^c{}_{d\nu\mu}. \tag{1.80}$$

This means that, just as $g_{\mu\nu}$, the spacetime-indexed quantities $\Gamma^\rho{}_{\nu\mu}$, $T^\lambda{}_{\mu\nu}$ and $R^\rho{}_{\lambda\nu\mu}$ are invariant under a local Lorentz transformation.

Comment 1.10 Let us register one more point. Suppose the members of a tetrad basis $\{h_a\}$ with commutation rule

$$[h_a, h_b] = f^c{}_{ab} h_c$$

are Lorentz-transformed:

$$h'^a = \Lambda^a{}_b(x) h^b.$$

Then, in order to keep the commutation rule in the form

$$[h'_a, h'_b] = f'^c{}_{ab} h'_c, \tag{1.81}$$

the anholonomy coefficients must transform in a very special way:

$$f'^c{}_{ab} = \Lambda^c{}_d(x) f^d{}_{ef} \Lambda^e{}_a(x) \Lambda^f{}_b(x) + \Lambda^c{}_d(x) [\Lambda^e{}_a(x) h_e \Lambda^d{}_b(x) - \Lambda^e{}_b(x) h_e \Lambda^d{}_a(x)]. \tag{1.82}$$

The last two, derivative terms vanish for global (independent of point, $\Lambda^c{}_b(x) \equiv \Lambda^c{}_b$) Lorentz transformations. The anholonomy coefficients are, in that case, just covariant:

$$f'^c{}_{ab} = \Lambda^c{}_d f^d{}_{ef} \Lambda^e{}_a \Lambda^f{}_b. \tag{1.83}$$

The derivative terms are, however, essential to compensate the behavior of the two connection terms in (1.59), keeping torsion covariant under local Lorentz transformations.

1.5 Bianchi Identities

Given a Lorentz connection $A^a{}_{b\mu}$, its torsion and curvature tensors satisfy two identities, called Bianchi identities. There is an identity for torsion,

$$\mathscr{D}_\nu T^a{}_{\rho\mu} + \mathscr{D}_\mu T^a{}_{\nu\rho} + \mathscr{D}_\rho T^a{}_{\mu\nu} = R^a{}_{\rho\mu\nu} + R^a{}_{\nu\rho\mu} + R^a{}_{\mu\nu\rho}, \tag{1.84}$$

usually called *first* Bianchi identity, and an identity for curvature,

$$\mathscr{D}_\nu R^a{}_{b\rho\mu} + \mathscr{D}_\mu R^a{}_{b\nu\rho} + \mathscr{D}_\rho R^a{}_{b\mu\nu} = 0, \tag{1.85}$$

usually called *second* Bianchi identity. Recall that the Fock-Ivanenko derivatives $\mathscr{D}_\nu$ act on algebraic indices only. Using relations (1.53) and (1.54), the Bianchi identity for torsion can be rewritten in the form

$$\begin{aligned} \nabla_\nu T^\lambda{}_{\rho\mu} + \nabla_\mu T^\lambda{}_{\nu\rho} + \nabla_\rho T^\lambda{}_{\mu\nu} &= R^\lambda{}_{\rho\mu\nu} + R^\lambda{}_{\nu\rho\mu} + R^\lambda{}_{\mu\nu\rho} \\ &+ T^\lambda{}_{\rho\sigma} T^\sigma{}_{\mu\nu} + T^\lambda{}_{\nu\sigma} T^\sigma{}_{\rho\mu} + T^\lambda{}_{\mu\sigma} T^\sigma{}_{\nu\rho}. \end{aligned} \tag{1.86}$$

In a similar fashion, the Bianchi identity for curvature becomes

$$\nabla_\nu R^\lambda{}_{\sigma\rho\mu} + \nabla_\mu R^\lambda{}_{\sigma\nu\rho} + \nabla_\rho R^\lambda{}_{\sigma\mu\nu}$$
$$= R^\lambda{}_{\sigma\mu\theta} T^\theta{}_{\nu\rho} + R^\lambda{}_{\sigma\nu\theta} T^\theta{}_{\rho\mu} + R^\lambda{}_{\sigma\rho\theta} T^\theta{}_{\mu\nu}. \tag{1.87}$$

The covariant derivative ∇_ν, it is worth recalling, acts on spacetime indices.

In the specific case of General Relativity, where the relevant connection is the zero-torsion Levi-Civita connection $\overset{\circ}{\Gamma}{}^\lambda{}_{\mu\nu}$, the Bianchi identities reduce to the more familiar expressions

$$\overset{\circ}{R}{}^\lambda{}_{\rho\mu\nu} + \overset{\circ}{R}{}^\lambda{}_{\nu\rho\mu} + \overset{\circ}{R}{}^\lambda{}_{\mu\nu\rho} = 0 \tag{1.88}$$

and

$$\overset{\circ}{\nabla}_\nu \overset{\circ}{R}{}^\lambda{}_{\sigma\rho\mu} + \overset{\circ}{\nabla}_\mu \overset{\circ}{R}{}^\lambda{}_{\sigma\nu\rho} + \overset{\circ}{\nabla}_\rho \overset{\circ}{R}{}^\lambda{}_{\sigma\mu\nu} = 0. \tag{1.89}$$

The cyclic symmetry (1.88) of the curvature tensor is, thus, a consequence of a vanishing torsion. It is a remarkable fact that, although torsion vanishes, the Bianchi identity for torsion remains in General Relativity—as the so-called cyclic identity.

Comment 1.11 Connections appearing in gauge theories of Yang-Mills type are differential forms with values in the Lie algebras of the unitary $SU(n)$ groups. Because the bundles underlying these theories are not soldered, torsion is not defined for these connections. In this kind of gauge theories, which includes Electromagnetism, there is only one Bianchi identity, that for curvature. This is a crucial difference between Yang-Mills theories, where torsion does not exist, and General Relativity, where torsion does exist but vanishes [see Comment 1.7].

1.6 Levi-Civita Symbol

The totally antisymmetric Levi-Civita (or Kronecker) symbol is defined by

$$\varepsilon^{\mu\nu\rho\sigma} = \begin{cases} +1 & \text{if } \mu\nu\rho\sigma \text{ is an even permutation of } 0123 \\ -1 & \text{if } \mu\nu\rho\sigma \text{ is an odd permutation of } 0123 \\ 0 & \text{otherwise,} \end{cases} \tag{1.90}$$

with the reference value assumed to be $\varepsilon^{0123} = 1$. It satisfies the contraction identity

$$\varepsilon^{\mu\nu\rho\sigma} \varepsilon_{\alpha\beta\gamma\sigma} = -\begin{vmatrix} \delta^\mu_\alpha & \delta^\mu_\beta & \delta^\mu_\gamma \\ \delta^\nu_\alpha & \delta^\nu_\beta & \delta^\nu_\gamma \\ \delta^\rho_\alpha & \delta^\rho_\beta & \delta^\rho_\gamma \end{vmatrix}. \tag{1.91}$$

The additional contractions follow

$$\varepsilon^{\mu\nu\rho\sigma} \varepsilon_{\alpha\beta\rho\sigma} = -2\left(\delta^\mu_\alpha \delta^\nu_\beta - \delta^\nu_\alpha \delta^\mu_\beta\right), \tag{1.92}$$

$$\varepsilon^{\mu\nu\rho\sigma} \varepsilon_{\alpha\nu\rho\sigma} = -6\delta^\mu_\alpha, \tag{1.93}$$

and

$$\varepsilon^{\mu\nu\rho\sigma} \varepsilon_{\mu\nu\rho\sigma} = -24. \tag{1.94}$$

Nevertheless, under a general coordinate transformation $\varepsilon^{\mu\nu\rho\sigma}$ does not behave as a tensor, but as a tensor density of weight -1. Considering that

$$h = \det(h^a{}_\mu) = \sqrt{-g} \tag{1.95}$$

is also a tensor density of weight -1, the quantity

$$h^{-1}\varepsilon^{\mu\nu\rho\sigma} \tag{1.96}$$

turns out to be an ordinary contravariant tensor (a tensor density of weight zero). Similarly, one can show that

$$\varepsilon_{\alpha\beta\gamma\delta} = \varepsilon^{\mu\nu\rho\sigma} g_{\mu\alpha} g_{\nu\beta} g_{\rho\gamma} g_{\sigma\delta} \tag{1.97}$$

is a tensor density of weight $+1$. Hence the quantity

$$h\varepsilon_{\mu\nu\rho\sigma} \tag{1.98}$$

turns out to be an ordinary covariant tensor.

1.7 Torsion Decomposition

The torsion tensor $T_{\lambda\mu\nu}$ can be decomposed into three components [17], each one irreducible under the global Lorentz group: there will be a vector part

$$\mathscr{V}_\mu = T^\nu{}_{\nu\mu}, \tag{1.99}$$

an axial part

$$\mathscr{A}^\mu = \tfrac{1}{6}\varepsilon^{\mu\nu\rho\sigma} T_{\nu\rho\sigma}, \tag{1.100}$$

and a purely tensor part

$$\mathscr{T}_{\lambda\mu\nu} = \tfrac{1}{2}(T_{\lambda\mu\nu} + T_{\mu\lambda\nu}) + \tfrac{1}{6}(g_{\nu\lambda}\mathscr{V}_\mu + g_{\nu\mu}\mathscr{V}_\lambda) - \tfrac{1}{3}g_{\lambda\mu}\mathscr{V}_\nu, \tag{1.101}$$

that is, a tensor with vanishing vector and axial parts. By construction, the latter is symmetric in the first two indices,

$$\mathscr{T}_{\lambda\mu\nu} = \mathscr{T}_{\mu\lambda\nu}, \tag{1.102}$$

and satisfies the cyclic symmetry

$$\mathscr{T}_{\lambda\mu\nu} + \mathscr{T}_{\nu\lambda\mu} + \mathscr{T}_{\mu\nu\lambda} = 0. \tag{1.103}$$

The parts so defined are called *vector torsion*, *axial torsion* and *pure tensor torsion*. With them as components, the whole tensor can be rewritten as

$$T_{\lambda\mu\nu} = \tfrac{2}{3}(\mathscr{T}_{\lambda\mu\nu} - \mathscr{T}_{\lambda\nu\mu}) + \tfrac{1}{3}(g_{\lambda\mu}\mathscr{V}_\nu - g_{\lambda\nu}\mathscr{V}_\mu) + \varepsilon_{\lambda\mu\nu\rho}\mathscr{A}^\rho. \tag{1.104}$$

The names in the decomposition above come from their behavior under space inversion P (parity transformation) and time reversal T. We call *vectors* those objects exhibiting the following responses to P and T transformations:

$$\mathscr{V}^\mu = (\mathscr{V}^0, \mathscr{V}^k) \quad \xrightarrow{\mathsf{P}} \quad (\mathscr{V}^0, -\mathscr{V}^k)$$

and

$$\mathscr{V}^\mu = (\mathscr{V}^0, \mathscr{V}^k) \quad \xrightarrow{\mathsf{T}} \quad (-\mathscr{V}^0, \mathscr{V}^k).$$

An *axial-vector*, on the other hand, transforms according to

$$\mathscr{A}^\mu = (\mathscr{A}^0, \mathscr{A}^k) \quad \xrightarrow{\mathsf{P}} \quad (-\mathscr{A}^0, \mathscr{A}^k)$$

and

$$\mathscr{A}^\mu = (\mathscr{A}^0, \mathscr{A}^k) \quad \xrightarrow{\mathsf{T}} \quad (\mathscr{A}^0, -\mathscr{A}^k).$$

The quantity $\mathscr{V}^\mu \mathscr{A}_\mu$, for example, is a pseudo-scalar under both P and T transformations,

$$\mathscr{V}^\mu \mathscr{A}_\mu \quad \xrightarrow{\mathsf{P}} \quad -\mathscr{V}^\mu \mathscr{A}_\mu \qquad \mathscr{V}^\mu \mathscr{A}_\mu \quad \xrightarrow{\mathsf{T}} \quad -\mathscr{V}^\mu \mathscr{A}_\mu,$$

and a scalar under a combined PT transformation:

$$\mathscr{V}^\mu \mathscr{A}_\mu \quad \xrightarrow{\mathsf{PT}} \quad \mathscr{V}^\mu \mathscr{A}_\mu.$$

Comment 1.12 For the gravitational interaction of spinors, the purely tensor piece is irrelevant: only the vector and the axial torsions appear in the Dirac equation [see Sect. 12.4.2]. This last fact will be used in Sect. 15.4 in the search for a dual-symmetric toy model for gravity.

References

1. Kobayashi, S., Nomizu, K.: Foundations of Differential Geometry, 2nd edn. Wiley-Interscience, New York (1996)
2. Aldrovandi, R., Pereira, J.G.: An Introduction to Geometrical Physics. World Scientific, Singapore (1995)
3. Zeeman, E.C.: Topology **6**, 161 (1967)
4. Hawking, S.W., King, A.R., McCarthy, P.J.: J. Math. Phys. **17**, 174 (1976)
5. Göbel, R.: J. Math. Phys. **17**, 845 (1976)
6. Malament, D.B.: J. Math. Phys. **18**, 1399 (1977)
7. Briginshaw, A.J.: Int. J. Theor. Phys. **19**, 329 (1980)
8. Fullwood, D.T.: J. Math. Phys. **33**, 2232 (1992)
9. Naber, G.L.: The Geometry of Minkowski Spacetime. Springer, New York (1992)
10. Gibbons, G.W.: Class. Quantum Gravity **10**, 575 (1993)
11. Fock, V.A., Ivanenko, D.: Z. Phys. **54**, 798 (1929)
12. Fock, V.A.: Z. Phys. **57**, 261 (1929)
13. Ramond, P.: Field Theory: A Modern Primer, 2nd edn. Addison-Wesley, Redwood (1989)
14. Dirac, P.A.M.: In: Kockel, B., Macke, W., Papapetrou, A. (eds.) Planck Festscrift. Deutscher Verlag der Wissenschaften, Berlin (1958)
15. Veltman, M.J.G.: Quantum theory of gravitation. In: Balian, R., Zinn-Justin, J. (eds.) Methods in Field Theory, Les Houches 1975. North-Holland, Amsterdam (1976)
16. Greub, W., Halperin, S., Vanstone, R.: Connections, Curvature, and Cohomology: Lie Groups, Principal Bundles, and Characteristic Classes. Academic Press, New York (1973)
17. Hayashi, K., Bregman, A.: Ann. Phys. **75**, 562 (1973)

Chapter 2
Lorentz Connections and Inertia

In Special Relativity, Lorentz connections represent inertial effects present in non-inertial frames. In these frames, any relativistic equation acquires a manifestly Lorentz covariant form due to the presence of the inertial connection. By virtue of the local equivalence between gravitation and inertia, these notions are essential for the study of the gravitational interaction.

2.1 Purely Inertial Connection

In Special Relativity, Lorentz connections represent inertial effects present in a given frame. In the class of inertial frames, for example, where these effects are absent, the Lorentz connection vanishes identically. Since this is the class most used in Field Theory, Lorentz connections usually do not show up in relativistic physics. Of course, as long as Physics is frame independent, it can be described in any class of frames. For the sake of simplicity, however, one always uses the class of inertial frames.

To see how an inertial Lorentz connection shows up, let us denote by $e^a{}_\mu$ a generic frame in Minkowski spacetime. The class of inertial (or holonomic) frames, defined by all frames for which $f'^c{}_{ab} = 0$, will be denoted by $e'^a{}_\mu$ [see Eq. (1.14) and around]. In a general coordinate system, the frames belonging to this class have the holonomic form

$$e'^a{}_\mu = \partial_\mu x'^a, \tag{2.1}$$

with x'^a a spacetime-dependent Lorentz vector: $x'^a = x'^a(x)$. The spacetime metric

$$\eta'_{\mu\nu} = e'^a{}_\mu e'^b{}_\nu \eta_{ab} \tag{2.2}$$

still represents the Minkowski metric, but in a general coordinate system. In the specific case of cartesian coordinates, the inertial frame assumes the form

$$e'^a{}_\mu = \delta^a_\mu \tag{2.3}$$

R. Aldrovandi, J.G. Pereira, *Teleparallel Gravity*, Fundamental Theories of Physics 173,
DOI 10.1007/978-94-007-5143-9_2, © Springer Science+Business Media Dordrecht 2013

and the spacetime metric $\eta'_{\mu\nu}$ is given by (1.16). Under a local Lorentz transformation,

$$x^a = \Lambda^a{}_b(x) x'^b, \tag{2.4}$$

the holonomic frame (2.1) transforms according to

$$e^a{}_\mu = \Lambda^a{}_b(x) e'^b{}_\mu. \tag{2.5}$$

As a simple computation shows, it has the explicit form

$$e^a{}_\mu = \partial_\mu x^a + \overset{\bullet}{A}{}^a{}_{b\mu} x^b \equiv \overset{\bullet}{\mathscr{D}}_\mu x^a, \tag{2.6}$$

where

$$\overset{\bullet}{A}{}^a{}_{b\mu} = \Lambda^a{}_e(x) \partial_\mu \Lambda_b{}^e(x) \tag{2.7}$$

is a Lorentz connection that represents the inertial effects present in the new frame. It is just the connection obtained from a Lorentz transformation of the vanishing spin connection $\overset{\bullet}{A}{}'^e{}_{d\mu} = 0$, as can be seen from Eq. (1.78):

$$\overset{\bullet}{A}{}^a{}_{b\mu} = \Lambda^a{}_e(x) \overset{\bullet}{A}{}'^e{}_{d\mu} \Lambda_b{}^d(x) + \Lambda^a{}_e(x) \partial_\mu \Lambda_b{}^e(x). \tag{2.8}$$

Starting from an inertial frame, in which $\overset{\bullet}{A}{}'^a{}_{b\mu} = 0$, different classes of frames are obtained by performing *local* (point-dependent) Lorentz transformations $\Lambda^a{}_b(x^\mu)$. Inside each class, the infinitely many frames are related through *global* (point-independent) Lorentz transformations, $\Lambda^a{}_b = \text{constant}$.

The inertial connection (2.7) is sometimes referred to as the Ricci coefficient of rotation [1]. Due to its presence, the transformed frame $e^a{}_\mu$ is no longer holonomic. In fact, its coefficient of anholonomy is given by

$$f^c{}_{ab} = -\left(\overset{\bullet}{A}{}^c{}_{ab} - \overset{\bullet}{A}{}^c{}_{ba}\right), \tag{2.9}$$

where we have used the identity $\overset{\bullet}{A}{}^a{}_{bc} = \overset{\bullet}{A}{}^a{}_{b\mu} e_c{}^\mu$. The inverse relation is

$$\overset{\bullet}{A}{}^a{}_{bc} = \tfrac{1}{2}\left(f_b{}^a{}_c + f_c{}^a{}_b - f^a{}_{bc}\right). \tag{2.10}$$

Of course, as a purely inertial connection, $\overset{\bullet}{A}{}^a{}_{b\mu}$ has vanishing curvature and torsion:

$$\overset{\bullet}{R}{}^a{}_{b\nu\mu} \equiv \partial_\nu \overset{\bullet}{A}{}^a{}_{b\mu} - \partial_\mu \overset{\bullet}{A}{}^a{}_{b\nu} + \overset{\bullet}{A}{}^a{}_{e\nu} \overset{\bullet}{A}{}^e{}_{b\mu} - \overset{\bullet}{A}{}^a{}_{e\mu} \overset{\bullet}{A}{}^e{}_{b\nu} = 0 \tag{2.11}$$

and

$$\overset{\bullet}{T}{}^a{}_{\nu\mu} \equiv \partial_\nu e^a{}_\mu - \partial_\mu e^a{}_\nu + \overset{\bullet}{A}{}^a{}_{e\nu} e^e{}_\mu - \overset{\bullet}{A}{}^a{}_{e\mu} e^e{}_\nu = 0. \tag{2.12}$$

2.2 Particle Equation of Motion

In the class of inertial frames $e'^a{}_\mu$, a free particle is described by the equation of motion

$$\frac{du'^a}{d\sigma} = 0, \tag{2.13}$$

with u'^a the particle four-velocity, and

$$d\sigma^2 = \eta_{\mu\nu}\, dx^\mu\, dx^\nu \tag{2.14}$$

the quadratic Minkowski invariant interval, with σ the proper time. In a anholonomic frame $e^a{}_\mu$, related to $e'^a{}_\mu$ by the local Lorentz transformation (2.5), the equation of motion assumes the manifestly covariant form

$$\frac{du^a}{d\sigma} + \overset{\bullet}{A}{}^a{}_{b\mu} u^b u^\mu = 0, \tag{2.15}$$

where

$$u^a = \Lambda^a{}_b(x) u'^b \tag{2.16}$$

is the Lorentz transformed four-velocity, and

$$u^\mu = u^a e_a{}^\mu \tag{2.17}$$

is the spacetime-indexed four-velocity, which has the usual holonomic form

$$u^\mu = \frac{dx^\mu}{d\sigma}. \tag{2.18}$$

It is important to remark that, since the anholonomy of the new frame $e^a{}_\mu$ is related to inertial effects only, the spacetime metric

$$\eta_{\mu\nu} = e^a{}_\mu e^b{}_\nu \eta_{ab}, \tag{2.19}$$

still represents the Minkowski metric, though in a general x-dependent form.

Comment 2.1 This allows a better characterization of the metrics defined by (1.27). Anholonomic basis fields related by Lorentz transformations define one same metric. On the other hand, anholonomic basis fields not related by Lorentz transformations define different metrics. This second case is what happens in the presence of gravitation. In fact, a tetrad whose anholonomy represents a gravitational field cannot be obtained through a Lorentz transformation from a tetrad whose anholonomy represents inertial effects only.

In terms of the holonomic four-velocity, the equation of motion (2.15) assumes the form

$$\frac{du^\rho}{d\sigma} + \overset{\bullet}{\gamma}{}^\rho{}_{\nu\mu} u^\nu u^\mu = 0, \tag{2.20}$$

where

$$\overset{\bullet}{\gamma}{}^\rho{}_{\nu\mu} = e_c{}^\rho \partial_\mu e^c{}_\nu + e_c{}^\rho \overset{\bullet}{A}{}^c{}_{b\mu} e^b{}_\nu \equiv e_c{}^\rho \overset{\bullet}{\mathcal{D}}_\mu e^c{}_\nu \tag{2.21}$$

is the spacetime-indexed version of the inertial spin connection $\overset{\bullet}{A}{}^a{}_{b\mu}$, obtained through contractions with the trivial tetrad $e^a{}_\mu$. Of course, since it has vanishing torsion, it is symmetric in the last two indices:

$$\overset{\bullet}{\gamma}{}^\rho{}_{\nu\mu} = \overset{\bullet}{\gamma}{}^\rho{}_{\mu\nu}. \tag{2.22}$$

The inverse relation is

$$\overset{\bullet}{A}{}^a{}_{b\mu} = e^a{}_\rho \partial_\mu e_b{}^\rho + e^a{}_\rho \overset{\bullet}{\gamma}{}^\rho{}_{\nu\mu} e_b{}^\nu \equiv e^a{}_\rho \overset{\bullet}{\nabla}_\mu e_b{}^\rho. \tag{2.23}$$

In an inertial frame $e'^a{}_\mu$, where $\stackrel{\bullet}{A}'^a{}_{b\mu} = 0$, we see from Eq. (2.21) that

$$\stackrel{\bullet}{\gamma}'^\rho{}_{\nu\mu} = e'_c{}^\rho \partial_\mu e'^c{}_\nu. \tag{2.24}$$

In cartesian coordinates, where $e'^a{}_\mu = \delta^a_\mu$, the connection $\stackrel{\bullet}{\gamma}'^\rho{}_{\nu\mu}$ vanishes and the equation of motion (2.20) assumes the usual form

$$\frac{du'^\rho}{d\sigma} = 0 \tag{2.25}$$

with $u'^\rho = e'^a{}_\mu u'^a$.

We have shown above how inertial and coordinate effects show up in the equation of motion of a free particle. Actually, this can be done for any relativistic equation. For example, in an inertial frame, and using cartesian coordinates, the sourceless Maxwell's equation reads

$$\partial_\mu F^{\mu\nu} = 0, \tag{2.26}$$

where

$$F^{\mu\nu} = \partial^\mu A^\nu - \partial^\nu A^\mu \tag{2.27}$$

is the field strength, with A^μ the electromagnetic potential. In a non-inertial frame, and considering a general coordinate system, it assumes the manifestly covariant form

$$\stackrel{\bullet}{\nabla}_\mu F^{\mu\nu} = 0, \tag{2.28}$$

with the field strength given now by

$$F^{\mu\nu} = \stackrel{\bullet}{\nabla}^\mu A^\nu - \stackrel{\bullet}{\nabla}^\nu A^\mu = \partial^\mu A^\nu - \partial^\nu A^\mu, \tag{2.29}$$

the last equality coming from the symmetry of $\stackrel{\bullet}{\gamma}^\rho{}_{\nu\mu}$ in the last two indices. And so on for any relativistic equation of Physics. Of course, for the sake of simplicity, one always uses inertial frames when dealing with field theory, where these effects are absent and the inertial connection (2.7) vanishes.

Comment 2.2 Inertial frames can only be defined in absence of gravitation. In the presence of gravitation, however, it is possible to define some generalizations. For example, one can introduce the notion of inertia-free frames. In these *global* frames, the inertial connection $\stackrel{\bullet}{A}^a{}_{b\nu}$ vanishes, and consequently their coefficients of anholonomy represent gravitation only, not inertia. This kind of frame can only be defined in the context of Teleparallel Gravity, and will be studied in more detail in Chap. 6. On the other hand, in the context of General Relativity, it is possible to define what is usually called a *locally inertial frame* [2]. It is a local frame in which the general relativistic spin connection $\stackrel{\circ}{A}^a{}_{b\nu}$ vanishes. In such local frame gravitation is exactly compensated by inertial effects, so that gravitation becomes undetectable at a given point. Of course, in order to produce an inertial effect that exactly compensates gravitation, this frame must be accelerated, and consequently cannot be inertial. Its name comes from the fact that, in this local frame the laws of Physics reduce to that of Special Relativity *as described from an inertial frame.*

2.3 Four-Acceleration and Parallel Transport

Let us consider now a general riemannian spacetime with metric

$$g_{\mu\nu} = \eta_{ab} h^a{}_\mu h^b{}_\nu. \tag{2.30}$$

A curve $\gamma(s)$ on this spacetime, parametrized by proper time s, with

$$ds^2 = g_{\mu\nu}\, dx^\mu\, dx^\nu, \tag{2.31}$$

will have as four-velocity the vector of components

$$u^\mu = \frac{dx^\mu}{ds}. \tag{2.32}$$

The corresponding four-acceleration cannot be given a covariant meaning without a connection—and each different connection $\Gamma^\rho{}_{\mu\nu}$ will define a different four-acceleration

$$a^\rho \equiv u^\nu \nabla_\nu u^\rho = \frac{du^\rho}{ds} + \Gamma^\rho{}_{\mu\nu} u^\mu u^\nu. \tag{2.33}$$

But acceleration must remain a measure of the velocity variation with *time*, and time appears in that formula as the proper time defined by metric $g_{\mu\nu}$. If acceleration is to keep a meaning, it is necessary that the same metric be considered all along the curve. In other words, the acceleration-defining connection must parallel-transport $g_{\mu\nu}$, satisfying the metricity condition (1.46).

Comment 2.3 Observe that, differently from the four-acceleration, the definition of the four-velocity does not require a connection. In fact, even defined with an ordinary derivative, the four-velocity u^μ turns out to be a four-vector. The reason is that x^μ is not a four-vector, but a set of four scalar functions $\gamma^\mu(s)$ parameterizing the curve γ. As such, its ordinary derivative turns out to be covariant. This is similar to what happens with a scalar field, whose ordinary four-derivative is (in this case) covariant.

As a^ρ is orthogonal to u^ρ, its vanishing means that the u^ρ keeps parallel to itself along the curve. This leads to the notion of parallel transport: we say that u^ρ is parallel-transported along γ when $a^\rho = 0$. Further, as every vector field is locally tangent to a curve (its local "integral curve"), a condition like

$$z^\mu \nabla_\mu u^\rho \equiv \nabla_z u^\rho = 0 \tag{2.34}$$

says that u^ρ is parallel-transported along the integral curve of z^ρ. The metric compatibility condition (1.46) implies that

$$z^\rho \nabla_\rho g_{\mu\nu} = 0 \tag{2.35}$$

for any vector field z^ρ, which is equivalent to say that the metric $g_{\mu\nu}$ is parallel-transported everywhere on spacetime. This is true, in particular, for the Levi-Civita connection (1.66), in which case we get

$$z^\rho \overset{\circ}{\nabla}_\rho g_{\mu\nu} = 0. \tag{2.36}$$

2.4 Inertial Effects

Let us consider now an observer attached to a particle moving along curve γ. An observer is abstractly conceived as a timelike worldline [3, 4]. More: notice that the four members of a tetrad are (pseudo-)orthogonal to each other. This means that one of them is timelike, and the other three are spacelike. As

$$\eta_{ab} = h_{a\nu} h_b{}^\nu, \tag{2.37}$$

then

$$h_{0\nu} h_0{}^\nu = \eta_{00} = +1,$$

so that, in our convention with $\boldsymbol{\eta} = \text{diag}(+1, -1, -1, -1)$, the member h_0 is timelike and has unit modulus. The remaining h_k $(k = 1, 2, 3)$ are spacelike. We then "attach" h_0 to the observer by identifying

$$u = h_0 = \frac{d}{ds}, \tag{2.38}$$

with components $u^\mu = h_0{}^\mu$. Of course, h_0 will be the observer velocity. The tetrad field, in this way, is made into a *reference frame*, with an observer attached to it.

Take now a general connection Γ and examine the corresponding frame acceleration

$$a^a_{(f)} \equiv h^a{}_\rho a^\rho_{(f)} = h^a{}_\rho \Gamma^\rho{}_{\mu\nu} h_0{}^\mu h_0{}^\nu + h^a{}_\rho h_0\big(h_0{}^\rho\big). \tag{2.39}$$

Comparing with the spin connection components,

$$A^a{}_{bc} \equiv h^a{}_\rho \nabla_{h_c} h_b{}^\rho = h^a{}_\rho \Gamma^\rho{}_{\mu\nu} h_b{}^\mu h_c{}^\nu + h^a{}_\rho h_c\big(h_b{}^\rho\big), \tag{2.40}$$

we see that

$$a^a_{(f)} = A^a{}_{00} \tag{2.41}$$

for whatever connection. As $A^a{}_{bc}$ is antisymmetric in the first two indices, only $a^k_{(f)}$ is different from zero. The definition

$$A^a{}_{bc} \equiv h^a{}_\rho \nabla_{h_c} h_b{}^\rho, \tag{2.42}$$

which in words is the covariant derivative of h_b along h_c, projected along h_a, provides a general interpretation for $A^a{}_{bc}$: it is a *generalized frame proper acceleration*.

These considerations give a new perception of the acceleration

$$a^k_{(f)} = \frac{du^k}{ds} + A^k{}_{bc} u^b u^c, \tag{2.43}$$

as seen from an accelerated frame. Besides the first, kinetic term, it includes contributions from the frame itself. It can be decomposed in the form

$$a^k_{(f)} = \frac{du^k}{ds} + a^k_{(f)} u^0 u^0 + 2[u \times \omega_{(f)}]^k u^0 + A^k{}_{ij} u^i u^j, \tag{2.44}$$

where

$$\omega^k_{(f)} = \tfrac{1}{2} \varepsilon^{kij} A_{ij0} \tag{2.45}$$

is the space tetrad members angular velocity. Some of the terms turning up in (2.44) can be easily recognized: piece

$$a^k_{lin} = a^k_{(f)} u^0 u^0 \tag{2.46}$$

represents the frame linear acceleration, whereas piece

$$a^k_{Cor} = 2[u \times \omega_{(f)}]^k u^0 \tag{2.47}$$

represents the Coriolis force. The last piece represents additional inertial effects present in the frame [5].

Comment 2.4 Another transport, distinct from parallel transport, can be introduced which absorbs the inertial effects. Applied on a four-vector z^ρ, it is given by the Fermi-Walker derivative:

$$\nabla^{FW}_u z^\rho = \nabla_u z^\rho + a_\nu u^\rho z^\nu - a^\rho u_\nu z^\nu.$$

In the specific case of the Levi-Civita connection of General Relativity, it assumes the form

$$\overset{\circ}{\nabla}{}^{FW}_u z^\rho = \overset{\circ}{\nabla}_u z^\rho + \overset{\circ}{a}_\nu u^\rho z^\nu - \overset{\circ}{a}{}^\rho u_\nu z^\nu.$$

Applied to the four-velocity, it reads

$$\overset{\circ}{\nabla}{}^{FW}_u u^\rho = \overset{\circ}{\nabla}_u u^\rho + \overset{\circ}{a}_\nu u^\rho u^\nu - \overset{\circ}{a}{}^\rho u_\nu u^\nu.$$

Using the identities $u_\nu u^\nu = 1$ and $\overset{\circ}{a}_\nu u^\nu = 0$, it reduces to

$$\overset{\circ}{\nabla}{}^{FW}_u u^\rho = \overset{\circ}{\nabla}_u u^\rho - \overset{\circ}{a}{}^\rho.$$

Since $\overset{\circ}{\nabla}_u u^\rho = \overset{\circ}{a}{}^\rho$, we see that it vanishes identically:

$$\overset{\circ}{\nabla}{}^{FW}_u u^\rho = 0.$$

In particular,

$$\overset{\circ}{\nabla}{}^{FW}_{h_0} h_0{}^\rho \equiv \overset{\circ}{\nabla}_{h_0} h_0{}^\rho - \overset{\circ}{a}{}^\rho = 0$$

implies that h_0, by the Fermi-Walker transport, is kept tangent along its own integral curve.

References

1. Misner, C.W., Thorne, K.S., Wheeler, J.A.: Gravitation. Freeman, San Francisco (1973)
2. Weinberg, S.: Gravitation and Cosmology. Wiley, New York (1972)
3. Synge, J.L.: Relativity: The General Theory. Wiley, New York (1960)
4. Hawking, S.W., Ellis, G.F.R.: The Large Scale Structure of Space-Time. Cambridge University Press, Cambridge (1973)
5. Singh, D., Papini, G.: Nuovo Cimento B **115**, 25 (2000). gr-qc/0007032

Chapter 3
Gauge Theories and Gravitation

We give here short and rough *résumés* on gauge models, which describe three of the four fundamental interactions of Nature—and of General Relativity, the standard theory for gravitation, which is not a gauge theory. No more than a "cast of characters", with only the main protagonists in each case. At the end, a discussion is presented on how the gauge paradigm might be applied to gravitation.

3.1 The Gauge Tenets

The gauge theories which successfully describe electromagnetic, electroweak and strong interactions are all concerned with point-dependent transformations occurring in "internal" spaces, that is, spaces unrelated to the (external) spacetime differentiable manifold. The point-dependence means merely that different transformations in internal space take place at different points of spacetime.

The forerunner of these theories is the Yang-Mills model, with the unitary group $SU(2)$ as the gauge group acting on isotopic spin (isospin) spaces of varied dimensions, each one carrying a different linear representation of $SU(2)$. The proton-neutron pair stays in a 2-dimensional space of doublets (p, n), the pions in a 3-dimensional space of triplets (π^1, π^2, π^3), and so on. Particles are represented by these multiplet fields, or by certain linear combinations of them. More precisely, once the fields are quantized, the particles turn up as their quanta: protons are the quanta of what we call "the proton field", pions are those of "the pion field", and so on. A particle which is insensitive to a certain gauge field is assigned to a singlet representation of the group, in which it will not respond to gauge transformations.

Fiber bundles are composite manifolds which encapsulate all the geometric aspects of these theories [1]. They are a combination of a base manifold (here, spacetime) and another space of interest (the gauge group, or any other space carrying one of its representations), built up with the strong proviso that the overall set of points constitute also a differentiable manifold. Given a point p on the base space,

R. Aldrovandi, J.G. Pereira, *Teleparallel Gravity*, Fundamental Theories of Physics 173, DOI 10.1007/978-94-007-5143-9_3, © Springer Science+Business Media Dordrecht 2013

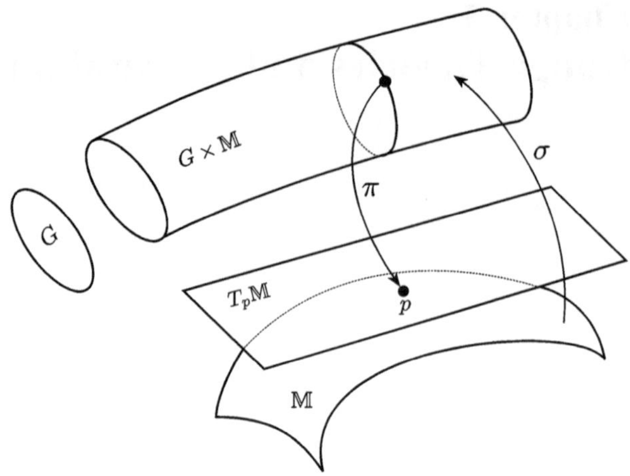

Fig. 3.1 Local view of a principal bundle

the bundle is locally (in a neighborhood of p) a direct product of both involved spaces.

Comment 3.1 With the circumference S^1 and the interval $(-1, +1)$ two simple but quite distinct bundles can be built: a cylinder, which is a global direct product, and a Möbius band, which is only locally a direct product.

A first bundle is constructed by attaching a copy of the gauge group G itself at each spacetime point p. Each fiber—space attached at each point of the base spacetime manifold—is itself a group, and in this case the bundle is said to be "principal" (see Fig. 3.1, where we have taken for base the Minkowski space $\mathbb{M}$). A "bundle projection" π takes all the points of the "fiber over p" into its corresponding base-space point p. And a converse "section" σ takes points on a neighborhood of p into a domain of the bundle manifold.

Other bundles, called "associated", are obtained by replacing the group by one of its linear representations. Source fields inhabit precisely the carrier vector spaces of such representations. They experience gauge transformations in the form

$$\Psi'^i(x) = U^i{}_j(x)\Psi^j(x), \tag{3.1}$$

where $U^i{}_j(x)$ [with $i, j, k = 1, 2, 3, \ldots, d =$ the *dimension of the representation*] are the entries of the matrix $U(x^\mu)$—the group element—representing the gauge transformation at the point p of coordinates x^μ. This is actually the way "sections" transform. Notice that to each principal bundle corresponds an infinity of associated bundles, one for each group representation. The principal bundle has not that name for nothing: theorems proved for it can afterwards be transferred to each one of its infinite associates.

3.1.1 Gauge Transformations

For an associated source field Ψ belonging to a given representation, the group element assumes the form $(A, B, C = 1, 2, 3, \ldots, n)$

$$U^i{}_j(x) = \left[\exp\left[\varepsilon^B(x) T_B\right]\right]^i{}_j, \tag{3.2}$$

where T_B are the transformation generators in the corresponding representation, $\varepsilon^B(x)$ are the group parameters, and n is the *dimension of the group*. Dropping the matrix indices, the gauge transformation (3.1) is written as

$$\Psi'(x) = \exp\left[\varepsilon^B(x) T_B\right] \Psi(x). \tag{3.3}$$

The corresponding infinitesimal transformation, obtained for

$$\left\|\varepsilon^C(x)\right\| \ll 1,$$

is given by

$$\delta\Psi(x) \equiv \Psi'(x) - \Psi(x) = \varepsilon^B(x) T_B \Psi(x). \tag{3.4}$$

The generators T_B satisfy the commutation relation

$$[T_B, T_C] = f^A{}_{BC} T_A, \tag{3.5}$$

where $f^A{}_{BC}$ are the structure constants of the group Lie algebra. We recall that the Lie algebra of a Lie group has this very special property, that a basis exists in which the structure coefficients for the generator commutators are constants. The adjoint representation, whose generators we denote by J_B, is given by $n \times n$ matrices with the very structure constants for entries:

$$(J_B)^A{}_C = f^A{}_{BC}. \tag{3.6}$$

3.1.2 Gauge Potential and Field Strength

The gauge boson field, which mediates the interactions, is a 1-form assuming values in the Lie algebra of the gauge group:

$$A = T_C A^C{}_\mu dx^\mu. \tag{3.7}$$

From the general definition of covariant derivative [1]

$$D_\mu \Psi(x) = \partial_\mu \Psi(x) - A^B{}_\mu \frac{\delta\Psi(x)}{\delta\varepsilon^B(x)}, \tag{3.8}$$

together with the infinitesimal transformation (3.4), we find that the covariant derivative of the source (or matter) field $\Psi(x)$ is

$$D_\mu \Psi(x) = \partial_\mu \Psi(x) - A^B{}_\mu T_B \Psi(x), \tag{3.9}$$

where the 1-form $A_\mu = A^B{}_\mu T_B$ now takes values on the representation of the Lie algebra appropriated for the field $\Psi(x)$ on which it acts. The covariant derivative of an object has the same behavior as the object itself. For example, since the field $\Psi(x)$ transforms according to

$$\Psi'(x) = U(x)\Psi(x), \tag{3.10}$$

its covariant derivative must transform in the same way, that is,

$$D'_{\mu}\Psi'(x) = U(x)D_{\mu}\Psi(x). \tag{3.11}$$

From this condition we obtain the transformation law of the gauge potential,

$$A'_{\mu} = U(x)A_{\mu}U^{-1}(x) + U(x)\partial_{\mu}U^{-1}(x), \tag{3.12}$$

which shows that A_μ does not transform covariantly. This is so because A is actually a connection: the last, derivative term exhibits just the non-covariance necessary to compensate the non-covariance of the ordinary derivative—thereby making of (3.9) a covariant object. As previously said, usual derivatives are not covariant under point-dependent transformations. Notice that the compensating, derivative term in (3.12) is independent of the connection. Namely, it is the same for any connection, and it disappears if we take the difference between two connections. That difference is, consequently, a covariant object, a gauge tensor.

The gauge potential A_μ belongs to the adjoint representation of the gauge group. This means that the group element $U(x)$ in (3.12) must be written with the generators in the adjoint representation (3.6). The infinitesimal version of (3.12) is then

$$\delta A^{C}{}_{\mu} \equiv A'^{C}{}_{\mu} - A^{C}{}_{\mu} = -\left[\partial_{\mu}\varepsilon^{C}(x) + f^{C}{}_{BD}A^{B}{}_{\mu}\varepsilon^{D}(x)\right], \tag{3.13}$$

which can be rewritten as

$$\delta A^{C}{}_{\mu} = -D_{\mu}\varepsilon^{C}(x), \tag{3.14}$$

with D_μ the covariant derivative in the adjoint representation. On the other hand, the covariant derivative of A itself is its curvature 2-form,

$$F = \tfrac{1}{2}J_{A}F^{A}{}_{\mu\nu}\,dx^{\mu}\wedge dx^{\nu}, \tag{3.15}$$

whose components along generators J_A are just those of the field strength

$$F^{A}{}_{\mu\nu} = \partial_{\mu}A^{A}{}_{\nu} - \partial_{\nu}A^{A}{}_{\mu} + f^{A}{}_{BC}A^{B}{}_{\mu}A^{C}{}_{\nu}. \tag{3.16}$$

Under the gauge transformation (3.12), the field strength $F_{\mu\nu} = J_{A}F^{A}{}_{\mu\nu}$ changes according to

$$F'_{\mu\nu} = U(x)F_{\mu\nu}U^{-1}(x), \tag{3.17}$$

which means that F transforms covariantly. The corresponding infinitesimal transformation is

$$\delta F^{A}{}_{\mu\nu} \equiv F'^{A}{}_{\mu\nu} - F^{A}{}_{\mu\nu} = f^{A}{}_{BC}\varepsilon^{B}F^{C}{}_{\mu\nu}. \tag{3.18}$$

Comment 3.2 Seen from a non-inertial (or anholonomic) frame, the gauge field strength assumes, instead of (3.16), the form

$$F^{C}{}_{ab} = h_{a}\big(A^{C}{}_{b}\big) - h_{b}\big(A^{C}{}_{a}\big) + f^{C}{}_{DE}A^{D}{}_{a}A^{E}{}_{b} - f^{d}{}_{ab}A^{C}{}_{d}. \tag{3.19}$$

The last term comes from the frame anholonomicity,

$$[h_c, h_d] = f^e{}_{cd} h_e,$$

and is linear in the connection. It is important not to confuse the structure constants of the group Lie algebra $f^C{}_{DE}$ with the coefficient of anholonomy $f^d{}_{ab}$ of the frame.

It is a good point to have in mind the well-known case of electromagnetism, which is a gauge theory for the gauge group $U(1)$. As a manifold, $U(1)$ is the 1-dimensional sphere, just the circumference $\mathbb{S}^1$. In this abelian case, $f^A{}_{BC}$ vanishes. Furthermore, because the group dimension is $n = 1$, the algebraic indices are usually omitted. In this case, the gauge transformation (3.13) reduces to

$$\delta A_\mu = -\partial_\mu \varepsilon(x), \tag{3.20}$$

and the field strength (3.16) assumes the usual expression for the Maxwell tensor

$$F_{\mu\nu} = \partial_\mu A_\nu - \partial_\nu A_\mu. \tag{3.21}$$

The observable, measurable field is $F_{\mu\nu}$, though the Aharonov-Bohm effect shows that some effects of A_μ can be measurable at the quantum level. The potential A_μ is, however, the fundamental field: the photon is the quantum of field A_μ. Electromagnetism has been the historical prototype of a gauge theory, even though its simplicity left many aspects of the gauge paradigm unnoticed. It has, for example, inspired the minimal coupling prescription (3.9)—though in its abelian, one-dimensional simplicity T_B is single and can be taken as a constant, say $T = \sqrt{\alpha}$, with

$$\alpha = \frac{q^2}{\hbar c} \simeq 1/137 \tag{3.22}$$

the fine-structure constant, the dimensionless coupling constant of the electromagnetic interaction (in gaussian units).

3.1.3 Field Equations

From the field strength definition (3.16) follows the identity

$$D_\rho F^A{}_{\mu\nu} + D_\nu F^A{}_{\rho\mu} + D_\mu F^A{}_{\nu\rho} = 0. \tag{3.23}$$

This is the *Bianchi identity*, which generalizes to the non-abelian case the first pair of Maxwell's equations,

$$\partial_\rho F_{\mu\nu} + \partial_\nu F_{\rho\mu} + \partial_\mu F_{\nu\rho} = 0. \tag{3.24}$$

Remember that these equations are not dynamical, in the sense that they do not follow from any lagrangian. The dynamical equations follow from the gauge lagrangian

$$\mathscr{L} = -\tfrac{1}{4}\gamma_{AB} F^A{}_{\mu\nu} F^{B\mu\nu}, \tag{3.25}$$

where

$$\gamma_{AB} = \text{Tr}(J_A J_B) = f^C{}_{AD} f^D{}_{BC} \tag{3.26}$$

is the Cartan-Killing metric, which is used to raise and lower internal indices. Since this metric can be defined only for semisimple groups (those with no invariant abelian subgroups), lagrangian (3.25) is only defined for these groups [2, 3]. For non-semisimple groups, such as the Poincaré group, the Cartan-Killing bilinear form γ_{AB} is not a metric—it is degenerate, and consequently not invertible.

Comment 3.3 Some abelian (non-semisimple) groups admit the construction of a lagrangian of the form (3.25), because some other gauge-invariant metric happens to exist. The most prominent example has been mentioned just above: electromagnetism, a gauge theory for the one-dimensional group $U(1)$. Its gauge lagrangian is constructed using, not the Cartan-Killing metric, but a different gauge-invariant metric, usually chosen as the constant $\gamma = 1$. Another example is Teleparallel Gravity, which corresponds to a gauge theory for the abelian translation group. In this case, the chosen metric is the Minkowski metric: $\gamma_{ab} = \eta_{ab}$. A more detailed discussion of this point will be presented in Chap. 9.

Let us then consider the lagrangian

$$\mathscr{L} = \mathscr{L}_s - \tfrac{1}{4} F^A{}_{\mu\nu} F_A{}^{\mu\nu}, \tag{3.27}$$

where the source lagrangian $\mathscr{L}_s = \mathscr{L}_s[\Psi, D_\mu \Psi]$ is obtained from the free lagrangian for the source multiplet field $\Psi = \{\Psi_j\}$ by the *minimal coupling prescription*: ordinary derivatives ∂_μ are replaced by covariant derivatives D_μ, given in this case by (3.9). Making use of the cyclic property

$$f_{ABC} = f_{CAB} = f_{BCA}, \tag{3.28}$$

characteristic of semisimple groups, the field equations that follow from the lagrangian (3.27) are the Yang-Mills equations

$$\partial_\mu F^{A\mu\nu} + f^A{}_{BC} A^B{}_\mu F^{C\mu\nu} = J^{A\nu}, \tag{3.29}$$

where

$$J^{A\nu} = -\frac{\partial \mathscr{L}_s}{\partial A_{A\nu}} \tag{3.30}$$

is the source current. These are the equations which generalize the second, dynamical pair of Maxwell's equations

$$\partial_\mu F^{\mu\nu} = J^\nu. \tag{3.31}$$

Notice that knowledge of the group structure constants is enough to write down the Yang-Mills equations (3.29), which can also be written in the form

$$D_\mu F^{A\mu\nu} = J^{A\nu}. \tag{3.32}$$

The non-abelian character brings forward non-linearity: the gauge field interacts with itself in the left-hand side of the field equation (3.29). The abelian photon does not carry electromagnetic charge, and consequently does not interact with itself. The gluon of strong interactions, however, as described by Chromodynamics, carries the strong $SU(3)_{\text{color}}$ charge. The second term in the left-hand side of (3.29) is just (minus) the self-current $j^{A\nu}$, the current carried by the gauge fields themselves:

$$j^{A\nu} = -f^A{}_{BC} A^B{}_\mu F^{C\mu\nu}. \tag{3.33}$$

From Noether's theorem, it can be verified that the source current is covariantly conserved:

$$D_\nu J^{A\nu} = 0. \tag{3.34}$$

Actually, this vanishing *covariant* divergence leads to no real conservation law—to something which is conserved in time. It is only a constraint on the source currents, usually called Noether identity. On the other hand, the field equation does lead to a true conservation: due to the antisymmetry of $F^{A\mu\nu}$ in the spacetime indices, we see that

$$\partial_\nu \left(J^{A\nu} + j^{A\nu} \right) = 0. \tag{3.35}$$

This equation says that the *total* current—source plus gauge field—is conserved. Notice, however, that the self-current $j^{A\nu}$ is not, by itself, covariant under gauge transformations. As a matter of fact, this is a consistency requirement: since the usual derivative is not covariant, the conserved current cannot be covariant either in such a way that the conservation law as a whole is covariant, and consequently physically meaningful.

Comment 3.4 The "problem" of the pseudo-current $j^{A\nu}$, which is not covariant under gauge transformations, is quite similar to the "problem" of the gravitational energy-momentum current, which is not covariant under general coordinate transformations. This question will be discussed in some more detail in Chap. 10.

3.1.4 Classical Equations of Motion

In the presence of an electromagnetic field, the Minkowski spacetime equation of motion of a test particle of mass m and electric charge q is described by the Lorentz force law

$$\frac{du^\mu}{d\sigma} = \frac{q}{mc^2} F^\mu{}_\nu u^\nu. \tag{3.36}$$

Generalization to a particle with a gauge charge q^A leads to

$$\frac{du^\mu}{d\sigma} = \gamma_{AB} \frac{q^A}{mc^2} F^{B\mu}{}_\nu u^\nu. \tag{3.37}$$

The charge itself will obey Wong's equation [4]

$$D_u q^A \equiv \frac{dq^A}{d\sigma} + A^B{}_\nu u^\nu (J_B)^A{}_C q^C = 0, \tag{3.38}$$

with $(J_B)^A{}_C = f^A{}_{BC}$ the generators written in the adjoint representation. This describes an "internal precession", which leads to

$$q_A D_u q^A = 0, \tag{3.39}$$

so that

$$\boldsymbol{q}^2 = \gamma_{AB} q^A q^B \tag{3.40}$$

is a characteristic invariant of the representation which is kept covariantly constant.

3.1.5 Duality Symmetry and Beyond

Gauge theories present a special, fundamental property: the *sourceless* dynamic equation

$$\partial_\mu F^{A\mu\nu} + f^A{}_{BC} A^B{}_\mu F^{C\mu\nu} = 0 \tag{3.41}$$

is just the geometrical identity (3.23) written for the dual of the field strength,

$$\star F^A{}_{\mu\nu} = \tfrac{1}{2}\varepsilon_{\mu\nu\rho\sigma} F^{A\rho\sigma}, \tag{3.42}$$

here taken on Minkowski spacetime. This is a remarkable property: if we know the geometry, we can obtain the dynamics, and vice-versa. This *duality symmetry* of the gauge field is very important for its quantization—it is, together with conformal symmetry, one of the attributes which make gauge theories *renormalizable*. Renormalizability is, in almost all cases, checked in perturbation theory, the perturbation parameter being the coupling constant, which is usually dimensionless. A well known example is the fine-structure constant α of Eq. (3.22), the coupling constant of the electromagnetic interaction. This means that, order by order, they multiply (Feynman) integrals of the same dimension, which allow all divergences to be eliminated.

Comment 3.5 In gravitation, as described by General Relativity, the coupling constant is $8\pi G/c^4$, with dimension $M^{-1}L^{-1}T^2$. If we try to quantize it, the Feynman integrals have, at each order, to compensate for these dimensions. It turns out that they become more and more divergent [5]. The origin of this difference is that generic Noether current densities have dimension MT^{-2}. The energy-momentum density, the source current for gravitation, has "abnormal" dimension $ML^{-1}T^{-2}$. The reason for this difference is that the energy-momentum conservation is related to the invariance of the lagrangians under spacetime translations. The point is that, unlike most gauge transformations, whose generators are dimensionless, the translation generator has dimension L^{-1}.

We have said that elementary particles appear in field theory as the quanta of the fundamental fields. Relativistic fields are actually sets of infinitely many degrees of freedom, one degree for each point of spacetime. In modern phenomenology, particles come usually before: they are first detected in Nature, and then a field is attributed to each of them. When a field is found beforehand (as in the electromagnetic case), a particle is identified to its quantum—provided the theory is quantizable. Models are built up by attributing particles to multiplets, and then collecting such multiplets into fields. These are, roughly speaking, the *source fields*. Interactions are then ascribed to *mediating fields*—here, just the gauge potentials. To the mediating fields are attributed new particles, the gauge bosons: photons for electromagnetism, gluons for the strong interactions regulated by Chromodynamics, the Z^0 and the $W^{(\pm)}$ for the weak sector of the electroweak interactions ruled by the group $SU(2) \otimes U(1)$. Non-renormalizability of a theory (as seems to be the case of gravitation, as described by General Relativity) jeopardizes its quantization—which becomes meaningless. We usually do speak of a "graviton" but, as long as General Relativity remains non-renormalizable, this is a mere convenience of language.

Gauge fields (or the particles they represent) are massless; there is no mass term in lagrangian (3.25). Actually, a mass term would violate gauge symmetry. Nevertheless, the afore mentioned mediating bosons Z^0 and $W^{(\pm)}$ do have masses, actually large masses as far as elementary particles are concerned. Such masses are theoretically obtained by the only known process which breaks the gauge symmetry while preserving renormalizability, the so-called *spontaneous symmetry breaking* (a name stemming from its original inspiration in superconductivity): a complex scalar field ϕ, which interacts with itself via a $\lambda\phi^4$ potential, is added to the gauge lagrangian [6]. The ensuing Hamiltonian exhibits a minimum value (vacuum) which stands below zero, and corresponds to a state which is degenerate. In other words, this vacuum state is multiple—one can pass from one state of minimum energy to another by a gauge transformation. It is necessary to choose one of these states as a fundamental state in order to build the higher energy states. This choice breaks the symmetry and induces a change of field variables. The original $SU(2) \otimes U(1)$ gauge fields are no more the physical fields. The two components of the added complex field compose with the original gauge potentials to produce the neutral Z^0 boson, the electrically charged $W^{(\pm)}$ bosons, and a residual field, the Higgs boson. With this special mechanism, the gauge symmetry is preserved as a "hidden" symmetry: it remains behind the scene, but holds for non-physical fields. Thus the massive, physically observed mediating fields come from a redefinition of the degrees of freedom.

3.2 General Relativity

Consider the set of all linear bases exchanged by the linear group $GL(4, \mathbb{R})$, or the set of tetrad frames exchanged by its Lorentz subgroup. Take a particular point p on the manifold $\mathbb{R}$ and choose one particular basis on the vector space $T_p\mathbb{R}$ to start with. Change then to any other: every other basis can be attained by a transformation which is a group member; and to each group element will correspond one basis obtained from the initial one. In consequence, the bundle of bases is the same as the bundle with one copy of the group at each point p—it is a principal bundle. It is one of the many miracles of tetrads that, being invertible, they determine the Lorentz transformation relating them [see Eq. (1.75)].

The principal bundle of frames puts all the geometry of General Relativity in a nutshell. A diagram is given in Fig. 3.2. The tangent bundle, with a tangent space

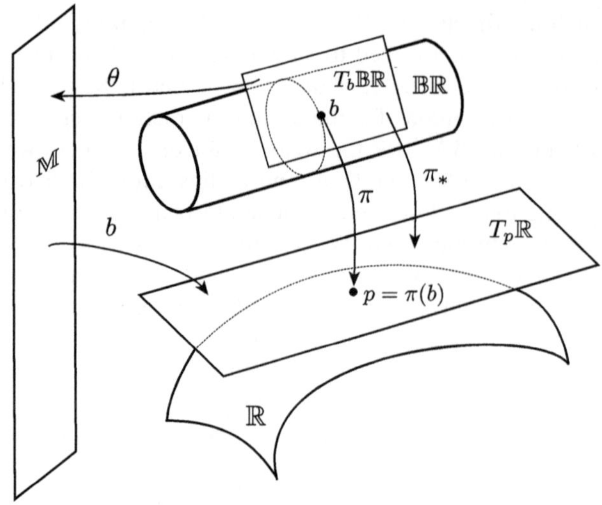

Fig. 3.2 Diagram of the frame bundle

$T_p\mathbb{R}$ attached to each point p of the riemannian spacetime $\mathbb{R}$, is only one of its associates. Tensor bundles, and spinor bundles, are others. In the bundle $\mathbb{BR}$ of bases on $\mathbb{R}$, the whole set of frames on $T_p\mathbb{R}$ is "attached" to point $p \in \mathbb{R}$. Its more intimate relation to the spacetime differentiable manifold structure makes it different from the corresponding gauge principal bundle pictured in Fig. 3.1. As repeatedly said, the main difference is the presence of soldering—with the ensuing appearance of tetrads and torsion.

Comment 3.6 The parallelizable manifolds mentioned in Comment 1.3 have a simple definition in terms of bundles: their tangent bundle is trivial, that is, globally a direct product of the typical tangent space (say, Minkowski space) by the base manifold (say, a riemannian spacetime). This is a special case of a very general property: a bundle is a global direct product (fiber × base) *iff* there exists a global section.

Consider a particular basis b on $T_p\mathbb{R}$: it will be a point on $\mathbb{BR}$. This basis, and all its companions obtained from it by a basis transformation, are taken into point p by the bundle projection π. The solder 1-form θ relates each tangent space $T_b\mathbb{BR}$ *of the bundle of frames* to the Minkowski space $\mathbb{M}$ and—this is the main property—has, seen from each tetrad, just the components of that same tetrad. Form θ acts in a circuitous way: the mapping b—so called because it just represents the homonymous basis b—in the diagram, which is a vector-space isomorphism, takes $\mathbb{M}$ into $T_p\mathbb{R}$ and makes of $T_p\mathbb{R}$ a Minkowski space. The torsion tensor of a given connection is just the covariant derivative of the solder 1-form or, due to the mentioned property, the covariant derivative of the tetrad field [see Eq. (1.52)]. Torsion is simply non-existent in internal (non-soldered) gauge theories.

General Relativity conceives the gravitational interaction—which it describes with paramount success at the classical level—as a change in the geometry of spacetime itself [7]. Specifically, as a change from the Lorentz metric $\eta_{\mu\nu}$ of Minkowski space into a riemannian metric $g_{\mu\nu}$. This new metric plays the role of basic field, and is in principle defined everywhere. Derivatives compatible with this overall presence of the same metric must preserve it, must parallel-transport it everywhere [see Sect. 2.3]. Of all such Lorentz connections preserving $g_{\mu\nu}$, the most natural choice is to pick up the Christoffel, or Levi-Civita connection

$$\overset{\circ}{\Gamma}{}^{\sigma}{}_{\mu\nu} = \tfrac{1}{2} g^{\sigma\rho}(\partial_\mu g_{\rho\nu} + \partial_\nu g_{\rho\mu} - \partial_\rho g_{\mu\nu}), \tag{3.43}$$

which is a connection determined solely by the ten components of the metric tensor $g_{\mu\nu}$. It is the only metric-preserving connection with vanishing torsion, a magnitude which is then found not to play any role in the general-relativistic description of the gravitational interaction [see Comment 1.7 though]. Its curvature, on the other hand, given by the Riemann tensor

$$\overset{\circ}{R}{}^{\rho}{}_{\lambda\mu\nu} = \partial_\mu \overset{\circ}{\Gamma}{}^{\rho}{}_{\lambda\nu} - \partial_\nu \overset{\circ}{\Gamma}{}^{\rho}{}_{\lambda\mu} + \overset{\circ}{\Gamma}{}^{\rho}{}_{\eta\mu} \overset{\circ}{\Gamma}{}^{\eta}{}_{\lambda\nu} - \overset{\circ}{\Gamma}{}^{\rho}{}_{\eta\nu} \overset{\circ}{\Gamma}{}^{\eta}{}_{\lambda\mu}, \tag{3.44}$$

represents the fundamental field of the theory: gravitation is present whenever at least one of its components is non-vanishing. The Ricci tensor is a symmetric second-order tensor defined as the contraction

$$\overset{\circ}{R}_{\mu\nu} = \overset{\circ}{R}{}^{\rho}{}_{\mu\rho\nu}, \tag{3.45}$$

and the scalar curvature is

$$\overset{\circ}{R} = g^{\mu\nu} \overset{\circ}{R}_{\mu\nu}. \tag{3.46}$$

Then comes the main point: through appropriate index contractions, the second Bianchi identity (1.89) can be rewritten in the form

$$\overset{\circ}{\nabla}_\nu \big(\overset{\circ}{R}{}^{\mu\nu} - \tfrac{1}{2} g^{\mu\nu} \overset{\circ}{R}\big) = 0. \tag{3.47}$$

On the other hand, from Noether's theorem, the invariance of a source (or matter) lagrangian $\mathcal{L}_s$ under general coordinate transformations yields the covariant conservation law

$$\overset{\circ}{\nabla}_\nu \Theta^{\mu\nu} = 0, \tag{3.48}$$

with

$$\Theta^{\mu\nu} = -\frac{1}{2\sqrt{-g}} \frac{\delta \mathcal{L}_s}{\delta g_{\mu\nu}} \tag{3.49}$$

the symmetric source energy-momentum tensor, modified by the presence of gravitation. It is thus natural to write

$$\overset{\circ}{R}{}^{\mu\nu} - \tfrac{1}{2} \overset{\circ}{R} g^{\mu\nu} = k \Theta^{\mu\nu}, \tag{3.50}$$

where k is some constant. Making the correspondence with Newton's law in the static weak-field limit, one determines $k = 8\pi G/c^4$ and the field equation, Einstein's equation, comes out as

$$\overset{\circ}{R}{}^{\mu\nu} - \tfrac{1}{2} \overset{\circ}{R} g^{\mu\nu} = \frac{8\pi G}{c^4} \Theta^{\mu\nu}. \tag{3.51}$$

This equation can be obtained from the lagrangian

$$\mathcal{L} = \overset{\circ}{\mathcal{L}} + \mathcal{L}_s, \tag{3.52}$$

where

$$\overset{\circ}{\mathscr{L}} = -\frac{c^4}{16\pi G}\sqrt{-g}\overset{\circ}{R} \tag{3.53}$$

is the Einstein-Hilbert lagrangian, and $\mathscr{L}_s$ is the matter, or source field lagrangian.

It is important to remark that the Einstein-Hilbert lagrangian (3.53) involves, in addition to $g_{\mu\nu}$ and its first derivatives, also second derivatives of $g_{\mu\nu}$. As a matter of fact, it is impossible to construct an invariant in terms of the metric and its first derivatives only. However, because the term containing second derivatives of the metric is linear in those second derivatives, that term can be transformed, by using Gauss theorem, into a divergence term. Namely, it is possible to write

$$\int \sqrt{-g}\overset{\circ}{R}\, d^4x = \int \sqrt{-g}\overset{\circ}{\mathscr{L}}_1\, d^4x + \int \partial_\mu\big(\sqrt{-g}w^\mu\big)\, d^4x, \tag{3.54}$$

where $\overset{\circ}{\mathscr{L}}_1$ contains only the metric and its first derivatives, and w^μ is a four-vector. Although the right-hand side is invariant under general coordinate transformations, each one of its pieces are not invariant. This means that, by performing coordinate transformations, it is possible to find infinitely many pairs $(\overset{\circ}{\mathscr{L}}_1, w^\mu)$ which satisfy the above condition.

How does General Relativity describe gravitation? The curvature of the Levi-Civita connection gives rise to a geometric description of the gravitational interaction. To understand what such a *geometrical description* does mean, let us consider the motion of a point-like, structureless test particle in a gravitational field. In Minkowski spacetime, such a particle obeys the equation

$$u^\nu \partial_\nu u^\rho \equiv \frac{du^\rho}{d\sigma} = 0, \tag{3.55}$$

with $d\sigma$ the Minkowski invariant interval (2.14). To obtain the equation valid in the presence of gravitation, a rule turns up, which is reminiscent of the gauge prescription: the minimal coupling prescription. According to this prescription, all ordinary derivatives must be replaced by covariant derivatives. With Einstein's choice of the Levi-Civita connection, the gravitational coupling prescription reads

$$\partial_\nu \rightarrow \overset{\circ}{\nabla}_\nu. \tag{3.56}$$

By this rule, the free equation of motion (3.55) becomes the usual geodesic equation

$$u^\nu \overset{\circ}{\nabla}_\nu u^\rho \equiv \frac{du^\rho}{ds} + \overset{\circ}{\Gamma}{}^\rho{}_{\mu\nu} u^\mu u^\nu = 0, \tag{3.57}$$

where

$$ds^2 = g_{\mu\nu}\, dx^\mu\, dx^\nu \tag{3.58}$$

is the riemannian spacetime quadratic interval. This equation describes the motion of a test particle in the presence of a gravitational field. It says essentially that the four-acceleration of the particle vanishes:

$$\overset{\circ}{a}{}^\rho = 0. \tag{3.59}$$

This means that in General Relativity *there is no concept of gravitational force*. In this theory, the gravitational interaction is geometrized: the presence of gravitation produces a curvature in spacetime, and the gravitational interaction is described by letting particles to follow freely the spacetime curvature.

Any other gauge interaction would contribute with a force term to the right-hand side of the geodesic equation. For example, the motion of a test particle of mass m, electric charge q and four-velocity u^ρ in the presence of both an electromagnetic and a gravitational field is described by a Lorentz force law which generalizes (3.36) to

$$\frac{du^\rho}{ds} + \overset{\circ}{\Gamma}{}^\rho{}_{\mu\nu} u^\mu u^\nu = \frac{q}{mc^2} F^\rho{}_\nu u^\nu. \tag{3.60}$$

In the case of a particle with a gauge charge q^A, the equation of motion (3.37) would be generalized to

$$\frac{du^\rho}{ds} + \overset{\circ}{\Gamma}{}^\rho{}_{\mu\nu} u^\mu u^\nu = \gamma_{AB} \frac{q^A}{mc^2} F^{B\rho}{}_\nu u^\nu. \tag{3.61}$$

We see that, while a gauge interaction engenders a force in the right-hand side of the equation, gravitation remains as a left-hand side, geometric effect.

3.3 Gravitation and the Gauge Paradigm

General Relativity is not a gauge theory. It differs from gauge theories in many ways. The most relevant points are the following:

- The basic field of gauge theories, with respect to which variations are taken, is a connection, the gauge potential. In General Relativity, it is the metric.
- There is a connection in General Relativity, but it is not a fundamental field: given a metric, the Levi-Civita connection is immediately known. In addition, the Levi-Civita connection is neither a true gravitational variable in the sense of classical fields, nor a genuine gravitational connection. The reason is that it represents both gravitation and inertial effects [see Sect. 6.6].
- The gauge lagrangians are quadratic in the curvature. In General Relativity the Einstein-Hilbert lagrangian is linear in the curvature.
- There is in General Relativity no scalar written in terms of the metric and its first derivative only. In fact, in addition to the metric and its first derivative, the Einstein-Hilbert lagrangian depends also on the second derivative of the metric.
- Gauge interactions always appear as a force, while in General Relativity gravitation appears as a geometric effect—there is no gravitational force.
- General Relativity does not have a gauge group. Sometimes, the set of diffeomorphisms is considered to constitute the gauge group of gravitation. This is false, because any theory can be written in a diffeomorphic-covariant form. As a matter of fact, diffeomorphism is empty of dynamical meaning. Furthermore, it takes place in spacetime (the base space), not in the tangent space, the fiber of the tangent bundle.

The question then arises: would it be possible to describe gravitation in an alternative way, as a gauge theory? To seek an answer to this question, let us play the gauge game. Recall the case of electromagnetism, a gauge theory for the unitary group $U(1)$. The source of the electromagnetic field is the electric four-current. According to Noether's theorem [8], a crucial piece of the gauge approach, this current is conserved due to invariance of the source lagrangian under *global* transformations of the group $U(1)$. In order to recover this symmetry for the case of *local* transformations of the same group, it is necessary to introduce a connection assuming values in the Lie algebra of the $U(1)$ group. This connection represents the electromagnetic potential, and electromagnetism emerges as a gauge theory for $U(1)$.

Analogously, the source of gravitation is energy and momentum. From Noether's theorem, the energy-momentum tensor is conserved provided the source lagrangian is invariant under spacetime translations. If gravity is to be described by a gauge theory with energy-momentum as source, it must be a gauge theory for the translation group. This theory is just the Teleparallel Equivalent of General Relativity—or Teleparallel Gravity—the theory we shall study all along this book.

Comment 3.7 The name Teleparallel Gravity has in the past been used to refer to the three-parameter Hayashi-Shirafuji theory, which will be briefly discussed in Sect. 9.6. Here, however, we will use it as a synonymous of the Teleparallel Equivalent of General Relativity, a theory that emerges for a specific choice of those parameters.

References

1. Aldrovandi, R., Pereira, J.G.: An Introduction to Geometrical Physics. World Scientific, Singapore (1995)
2. Aldrovandi, R., Pereira, J.G.: Phys. Rev. D **33**, 2788 (1986)
3. Aldrovandi, R., Pereira, J.G.: J. Math. Phys. **29**, 1472 (1988)
4. Wong, S.K.: Nuovo Cimento A **65**, 689 (1970)
5. Veltman, M.J.G.: Quantum theory of gravitation. In: Balian, R., Zinn-Justin, J. (eds.) Methods in Field Theory, Les Houches 1975. North-Holland, Amsterdam (1976)
6. Itzykson, C., Zuber, J.B.: Quantum Field Theory. McGraw-Hill, New York (1980)
7. Fock, V.A.: The Theory of Space, Time and Gravitation, 2nd edn. Pergamon, New York (1964)
8. Konopleva, N.P., Popov, V.N.: Gauge Fields. Harwood, New York (1980)

Chapter 4
Fundamentals of Teleparallel Gravity

The foundations of Teleparallel Gravity, a gauge theory for the translations group, are presented. The geometrical setting is described, the translational gauge potential introduced, and the notion of gauge transformations explained. The translational coupling prescription is discussed, and the consequences for the spacetime metric examined. The field strength of the theory is shown to be the torsion tensor.

4.1 Some Historical Remarks

Soon after General Relativity was given its final presentation as a new theory for the gravitational field, an attempt to unify gravitation and electromagnetism was made by H. Weyl in 1918 [1]. His beautiful proposal did not succeed, but introduced for the first time the notions of *gauge transformations* and *gauge invariance*, and can be considered the seed of what is known today as gauge theory [2, 3]. Another attempt in the same direction was made by A. Einstein [4], about ten years later. This attempt was based on the mathematical structure of teleparallelism, also frequently referred to as distant, or absolute parallelism. Roughly speaking, the idea was the introduction of a tetrad, a field of orthonormal bases on the tangent spaces at each point of the four-dimensional spacetime. The tetrad has sixteen components, whereas the gravitational field, represented by the spacetime metric, has only ten. The six additional degrees of freedom of the tetrad were then supposed by Einstein to be related to the six components of the electromagnetic field. This attempt of unification did not succeed either, among other reasons because the additional six degrees of freedom of the tetrad are actually eliminated by the six-parameter local Lorentz invariance of the theory. Like Weyl's work, however, it introduced concepts that remain important to the present day.

Comment 4.1 A detailed historical account of the teleparallel-based Einstein's unification theory can be found in Ref. [5]. Some additional remarks on the history of teleparallelism, as well as on the notion of torsion itself, can be found in Ref. [6].

R. Aldrovandi, J.G. Pereira, *Teleparallel Gravity*, Fundamental Theories of Physics 173, DOI 10.1007/978-94-007-5143-9_4,

In the early nineteen-twenties, another attempt to unify gravitation and electromagnetism was made by Th. Kaluza [7] and O. Klein [8], giving rise to the so-called Kaluza-Klein theory. At the same time, É. Cartan developed a modification of General Relativity, in which spacetime was allowed to have curvature and torsion, and which became known as Einstein-Cartan theory [9, 10]. Cartan assumed the linear connection to be metric preserving, and obtained the gravitational field equations from a variational principle. Like in Einstein theory, energy and momentum were source of curvature, whereas some other quantity, which was later identified with spin, was the algebraic source of torsion. It is interesting to observe that, when Cartan published his theory, spin was not yet discovered.

Comment 4.2 The Kaluza-Klein theory, as well as its teleparallel version, will be studied in Chap. 16.

After this initial period, characterized mainly by unifying aspirations, the notion of teleparallelism experienced no new advances for nearly three decades. Only in the nineteen-sixties did Møller [11] revive Einstein's original idea, no more for unification purposes, but as an attempt to find a tensorial complex for the gravitational energy-momentum density. It is interesting to remark that Møller succeeded in finding a complex which was invariant under general coordinate transformations, though not invariant under local Lorentz transformations. Following this work, Pellegrini and Plebanski [12] found a lagrangian formulation for teleparallel gravity, a problem that Møller reconsidered later [13]. In 1967, Hayashi and Nakano [14] formulated a gauge theory for the translation group, which was further developed by Hayashi [15]. Then, in 1976, Cho proved that the teleparallel lagrangian—with the coefficient of anholonomy replacing torsion—was invariant (up to a divergence) under local Lorentz transformations, and equivalent to the Einstein-Hilbert lagrangian of General Relativity [16]. A few years later, Hayashi [17] pointed out the connection between this theory and teleparallelism, and an attempt to unify these two developments was made by Hayashi and Shirafuji [18] in 1979. According to this approach, General Relativity—a theory that involves only curvature—was supplemented by a kind of teleparallel gravity—a theory that involves only torsion, and presents three free parameters, proposed to be determined by experiment. This theory, called New General Relativity, represented a new way of including torsion in General Relativity, actually an alternative to the scheme previously provided by the Einstein-Cartan approach. Since then, many additional contributions to the theory have been made, culminating with the theory we know today as the Teleparallel Equivalent of General Relativity, or just Teleparallel Gravity. Of course, many aspects and new developments are yet to be explored. Its basic foundations, however, can be considered by now to be fully settled.

Comment 4.3 A glimpse of New General Relativity will be presented in Sect. 9.6. The Einstein-Cartan approach will be discussed in some detail in Chap. 17.

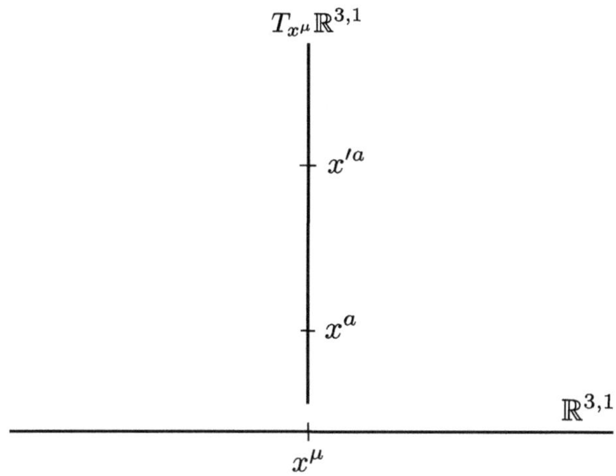

Fig. 4.1 Spacetime with the Minkowski tangent space at x^μ

4.2 Geometrical Setting

As said at the end of Sect. 3.3, Teleparallel Gravity corresponds to a gauge theory for the translation group. Its geometrical setting is the tangent bundle: at each point p of a general riemannian spacetime $\mathbb{R}^{3,1}$—the base space—there is "attached" a Minkowski tangent-space $\mathbb{M} = T_p\mathbb{R}^{3,1}$—the fiber—on which the gauge transformations take place. In Fig. 4.1, the tangent space at $p = \{x^\mu\}$ is indicated perpendicularly. A gauge transformation will be a point-dependent translation of the $T_p\mathbb{R}^{3,1}$ coordinates x^a,

$$x'^a = x^a + \varepsilon^a, \tag{4.1}$$

with $\varepsilon^a \equiv \varepsilon^a(x^\mu)$ the transformation parameters.

The generators of infinitesimal translations are the differential operators

$$P_a = \frac{\partial}{\partial x^a} \equiv \partial_a, \tag{4.2}$$

which satisfy the commutation relations

$$[P_a, P_b] = 0. \tag{4.3}$$

The corresponding infinitesimal transformation can then be written in the form

$$\delta x^a = \varepsilon^b P_b x^a. \tag{4.4}$$

Let us insist again on the fact that, due to the peculiar character of translations, any gauge theory including them will differ from the usual internal—Yang-Mills type—gauge models in many ways, the most significant being the presence of a tetrad field. The gauge bundle will then present the soldering property, and the "internal" and "external" sectors of the theory will be more closely linked to each other. Teleparallelism will be necessarily a non-standard gauge theory.

4.3 Gauge Transformations of Source Fields

Let us consider now a general source field Ψ. If V is an open set on spacetime, Ψ is represented by a *local section* Ψ_V of the fiber bundle, which is given by a differentiable application of the form [19]

$$\Psi_V : V \to \pi^{-1}(V), \tag{4.5}$$

where π is the bundle projection from the fiber into spacetime [see Figs. 3.1 and 3.2]. Now, a fiber bundle is always locally trivial, that is, $\pi^{-1}(V)$ is always diffeomorphic to $V \times F$, with F the fiber on which the gauge group acts:

$$\pi^{-1}(V) \sim V \times F. \tag{4.6}$$

This diffeomorphism, usually called a *local trivialization*, is given by

$$f_V : \pi^{-1}(V) \to V \times F. \tag{4.7}$$

For a fiber of dimension d, a local section Ψ_V is an application ($A = 1, \dots, d$)

$$x^\mu \to f_V^{-1}\big(x^\mu, x^A\big), \tag{4.8}$$

where the coordinate set $\{x^\mu\}$ indicates a point in spacetime and $\{x^A(x^\mu)\}$ indicates a point in the fiber over x^μ. Therefore, a general source field Ψ, defined as a section of an associate bundle, must depend on both coordinates x^μ and x^A:

$$\Psi = \Psi\big(x^\mu, x^A\big). \tag{4.9}$$

In the case of internal, Yang-Mills type gauge theories, the source field Ψ has a discrete set of components: it is a vector (a multiplet) in a fiber, which is the carrier space of the representation to which it belongs (see Sect. 3.1). For quantum-mechanical reasons, these representations must be unitary, and only compact groups have finite unitary representations. Thus, the fiber is a vector space of finite dimension, and the source fields are finite multiplets whose internal coordinates are the components. These components are distorted by a multi-component phase ε^A by a gauge transformation as in Eq. (3.3).

For the case of the non-compact translation group, unitarity is anyhow jeopardized. Each fiber is a copy of the whole Minkowski spacetime. The continuum of coordinates, now indicated $x^a(x^\mu)$, takes on the role of component indices in the above multiplets. The dependence of Ψ on $x^a(x^\mu)$ is written simply as

$$\Psi = \Psi\big(x^a(x^\mu)\big). \tag{4.10}$$

Under an infinitesimal tangent space translation, it transforms according to

$$\delta\Psi\big(x^a(x^\mu)\big) = \varepsilon^a \partial_a \Psi\big(x^a(x^\mu)\big). \tag{4.11}$$

This gives the functional change of Ψ at a fixed x^a and, of course, at fixed spacetime point x^μ—the typical transformation of gauge theories.

4.4 Gauge Coupling Prescription

It is a common practice in field theory to use inertial frames to describe the theory. Here, we will initially follow this practice, and will work in the inertial frame

$$e^a{}_\mu = \partial_\mu x^a. \tag{4.12}$$

We will later generalize the results to a non-inertial frame.

Comment 4.4 By an appropriate choice of the coordinates systems $\{x^a\}$ and $\{x^\mu\}$, it is always possible to re-write the above frame in the form

$$e^a{}_\mu = \delta^a_\mu. \tag{4.13}$$

Sometimes, for the sake of simplicity, we are going to use this specific form of an inertial frame. In general, however, we are going to use the general form (4.12), which is valid for any coordinate system.

4.4.1 The Electromagnetic Case as an Example

As an illustration of the general procedure, let us consider first the case of electromagnetism, a gauge theory for the unitary group $U(1)$. Under an infinitesimal gauge transformation with parameter ε, a general (complex) field Ψ changes according to

$$\delta\Psi = i\varepsilon\Psi. \tag{4.14}$$

For a global ($\varepsilon =$ constant) transformation, its ordinary derivative transforms covariantly:

$$\delta(\partial_\mu\Psi) = i\varepsilon(\partial_\mu\Psi). \tag{4.15}$$

For a local transformation with parameter $\varepsilon = \varepsilon(x)$, however, it does not transform covariantly anymore: actually,

$$\delta(\partial_\mu\Psi) = i\varepsilon(\partial_\mu\Psi) + i(\partial_\mu\varepsilon)\Psi. \tag{4.16}$$

In order to recover the covariance, it is necessary to introduce a gauge potential A_μ, which is a connection taking values at the Lie algebra of the gauge group $U(1)$. It is then easy to see that the new derivative

$$D_\mu\Psi = \partial_\mu\Psi + iA_\mu\Psi \tag{4.17}$$

transforms covariantly,

$$\delta(D_\mu\Psi) = i\varepsilon(D_\mu\Psi), \tag{4.18}$$

provided the gauge potential transforms according to

$$\delta A_\mu = -\partial_\mu\varepsilon. \tag{4.19}$$

The replacement

$$\partial_\mu\Psi \to D_\mu\Psi, \tag{4.20}$$

with $D_\mu\Psi$ the covariant derivative (4.17), defines the electromagnetic coupling prescription. Seen from the frame (4.12), it assumes the form

$$e^a{}_\mu \partial_a \Psi \to e^a{}_\mu D_a \Psi, \tag{4.21}$$

where

$$D_a \Psi = \partial_a \Psi + i A_a \Psi, \tag{4.22}$$

with

$$A_a = e_a{}^\mu A_\mu. \tag{4.23}$$

4.4.2 Translational Coupling Prescription

We use now the electromagnetic case as a guide to obtain the translational coupling prescription. As seen in Sect. 4.3, under an infinitesimal gauge translation, a general source field $\Psi = \Psi(x^a(x^\mu))$ transforms according to

$$\delta\Psi = \varepsilon^a \partial_a \Psi. \tag{4.24}$$

Like in the electromagnetic case, for a global (ε^a = constant) transformation, its ordinary derivative transforms covariantly:

$$\delta(\partial_\mu \Psi) = \varepsilon^a \partial_a (\partial_\mu \Psi). \tag{4.25}$$

For a local transformation with parameter $\varepsilon^a = \varepsilon^a(x)$, however, it does not transform covariantly anymore:

$$\delta(\partial_\mu \Psi) = \varepsilon^a \partial_a (\partial_\mu \Psi) + \left(\partial_\mu \varepsilon^a\right) \partial_a \Psi. \tag{4.26}$$

In order to recover the covariance, it is necessary to introduce a gauge potential B_μ, which in this case is a 1-form assuming values in the Lie algebra of the translation group, the gauge group of Teleparallel Gravity:

$$B_\mu = B^a{}_\mu P_a. \tag{4.27}$$

With this potential, one can define the derivative

$$h_\mu \Psi = \partial_\mu \Psi + B^a{}_\mu \partial_a \Psi, \tag{4.28}$$

which transforms covariantly under an infinitesimal gauge translation,

$$\delta(h_\mu \Psi) = \varepsilon^a \partial_a (h_\mu \Psi), \tag{4.29}$$

provided the gauge potential transforms according to

$$\delta B^a{}_\mu = -\partial_\mu \varepsilon^a. \tag{4.30}$$

The translational coupling prescription is then achieved by replacing

$$\partial_\mu \Psi \to h_\mu \Psi, \tag{4.31}$$

with $h_\mu\Psi$ the covariant derivative (4.28).

Comment 4.5 We remark that this covariant derivative can be obtained from the general definition of covariant derivative [19]

$$h_\mu = \partial_\mu + B^a{}_\mu \frac{\delta}{\delta \varepsilon^a}, \tag{4.32}$$

where

$$\frac{\delta}{\delta \varepsilon^a} = \frac{\partial}{\partial \varepsilon^a} - \partial_\rho \frac{\partial}{\partial(\partial_\rho \varepsilon^a)} + \cdots \tag{4.33}$$

is the Lagrange derivative with respect to the parameter ε^a. In fact, using the transformation (4.24), it is found to be just (4.28).

The covariant derivative (4.28) can be rewritten in the form

$$h_\mu \Psi = h^a{}_\mu \partial_a \Psi, \tag{4.34}$$

where

$$h^a{}_\mu = \partial_\mu x^a + B^a{}_\mu \tag{4.35}$$

is a non-trivial tetrad field. By non-trivial we mean a tetrad with

$$B^a{}_\mu \neq \partial_\mu \varepsilon^a, \tag{4.36}$$

otherwise it would be just a gauge transformation of (4.35). The translational coupling prescription (4.31) can then be rewritten in the form

$$e^a{}_\mu \partial_a \Psi \to h^a{}_\mu \partial_a \Psi, \tag{4.37}$$

from which we see that it just amounts to replace a trivial tetrad by a non-trivial one:

$$e^a{}_\mu \to h^a{}_\mu. \tag{4.38}$$

Concomitantly with this replacement, the spacetime metric changes according to

$$\eta_{\mu\nu} \to g_{\mu\nu}, \tag{4.39}$$

with the metrics given respectively by

$$\eta_{\mu\nu} = \eta_{ab} e^a{}_\mu e^b{}_\nu \quad \text{and} \quad g_{\mu\nu} = \eta_{ab} h^a{}_\mu h^b{}_\nu. \tag{4.40}$$

The change in spacetime metric is, therefore, a direct consequence of the translational coupling prescription.

4.4.3 Translational Coupling in a General Frame

Up to now, we have used a class of Lorentz frames in which no inertial effects are present. The equivalent expressions valid in a general Lorentz frame can be obtained by performing a local Lorentz transformation

$$x^a \to \Lambda^a{}_b(x) x^b, \tag{4.41}$$

under which

$$\Psi \to U(\Lambda)\Psi, \tag{4.42}$$

with $U(\Lambda)$ an element of the Lorentz group in the representation appropriate for the field Ψ. Considering that $B^a{}_\mu$ is a Lorentz vector in the tangent space index, that is,

$$B^a{}_\mu \to \Lambda^a{}_b(x) B^b{}_\mu, \tag{4.43}$$

it is immediate to see that the translational covariant derivative

$$h_\mu \Psi = h^a{}_\mu \partial_a \Psi \tag{4.44}$$

is now written with the tetrad

$$h^a{}_\mu = \partial_\mu x^a + \dot{A}^a{}_{b\mu} x^b + B^a{}_\mu, \tag{4.45}$$

where

$$\dot{A}^b{}_{c\mu} = \Lambda^b{}_d(x) \partial_\mu \Lambda_c{}^d(x) \tag{4.46}$$

is the purely inertial Lorentz connection introduced in Chap. 2. Tetrad (4.45) can be rewritten in the form

$$h^a{}_\mu = e^a{}_\mu + B^a{}_\mu, \tag{4.47}$$

where

$$e^a{}_\mu \equiv \dot{\mathcal{D}}_\mu x^a = \partial_\mu x^a + \dot{A}^a{}_{b\mu} x^b \tag{4.48}$$

is the trivial (that is, non-gravitational) part of the tetrad. In this class of frames, the gauge transformation of the translational potential $B^a{}_\mu$ is

$$\delta B^a{}_\mu = -\dot{\mathcal{D}}_\mu \varepsilon^a. \tag{4.49}$$

As can be easily verified, the tetrad is invariant under the gauge transformations (4.4) and (4.49):

$$\delta h^a{}_\mu = 0. \tag{4.50}$$

Comment 4.6 We remark that, since the generators $P_a = \partial_a$ are derivatives which act on matter fields $\Psi(x^a(x^\mu))$ through their tangent-space arguments x^a, every source field will respond equally to their action, and consequently will couple equally to the translational gauge potentials. All of them, therefore, will feel gravitation the same. This is the origin of the concept of *universality* according to Teleparallel Gravity.

4.5 Translational Field Strength

As in any gauge theory, the field strength of Teleparallel Gravity can be obtained from the commutation relation of gauge covariant derivatives. Using the translational covariant derivative (4.28), one can easily verify that

$$[h_\mu, h_\nu] = \dot{T}^a{}_{\mu\nu} P_a, \tag{4.51}$$

where

$$\dot{T}^a{}_{\mu\nu} = \partial_\mu B^a{}_\nu - \partial_\nu B^a{}_\mu + \dot{A}^a{}_{b\mu} B^b{}_\nu - \dot{A}^a{}_{b\nu} B^b{}_\mu \tag{4.52}$$

is the translational field strength. It can be rewritten in the form

$$\dot{T}^a{}_{\mu\nu} = \dot{\mathscr{D}}_\mu B^a{}_\nu - \dot{\mathscr{D}}_\nu B^a{}_\mu. \tag{4.53}$$

Adding the vanishing torsion

$$\dot{\mathscr{D}}_\mu(\dot{\mathscr{D}}_\nu x^a) - \dot{\mathscr{D}}_\nu(\dot{\mathscr{D}}_\mu x^a) \equiv [\dot{\mathscr{D}}_\mu, \dot{\mathscr{D}}_\nu] x^a = 0$$

to the right-hand side, it becomes

$$\dot{T}^a{}_{\mu\nu} = \dot{\mathscr{D}}_\mu(\dot{\mathscr{D}}_\nu x^a + B^a{}_\nu) - \dot{\mathscr{D}}_\nu(\dot{\mathscr{D}}_\mu x^a + B^a{}_\mu). \tag{4.54}$$

Remembering that

$$\dot{\mathscr{D}}_\mu x^a + B^a{}_\mu = h^a{}_\mu, \tag{4.55}$$

we see that the field strength of Teleparallel Gravity is nothing else, but torsion:

$$\dot{T}^a{}_{\mu\nu} = \dot{\mathscr{D}}_\mu h^a{}_\nu - \dot{\mathscr{D}}_\nu h^a{}_\mu. \tag{4.56}$$

Since the tetrad is gauge invariant, the field strength $\dot{T}^a{}_{\mu\nu}$ is also invariant under gauge transformations,

$$\dot{T}'^a{}_{\mu\nu} = \dot{T}^a{}_{\mu\nu}, \tag{4.57}$$

which is an expected result. In fact, because the generators of the adjoint representation are the coefficients of structure of the group taken as matrices, and considering that these coefficients vanish for abelian groups, fields belonging to the adjoint representations of abelian gauge theories will always be gauge invariant. This is precisely the case of the electromagnetic field strength $F_{\mu\nu}$, as well as of the torsion tensor.

4.6 Fundamental Fields

As a gauge field for the translation group, the gravitational field in Teleparallel Gravity is represented by the translational gauge potential $\boldsymbol{B}$, a 1-form assuming values in the Lie algebra of the translation group:

$$\boldsymbol{B} = B^a{}_\mu P_a \, dx^\mu. \tag{4.58}$$

The corresponding field strength is the torsion $\boldsymbol{T}$, a 2-form also assuming values in the Lie algebra of the translation group:

$$\boldsymbol{T} = \tfrac{1}{2} T^a{}_{\mu\nu} P_a \, dx^\mu \wedge dx^\nu. \tag{4.59}$$

The fundamental Lorentz connection of Teleparallel Gravity, on the other hand, is the purely inertial connection (4.46). This means that in this theory Lorentz connections keep the special-relativistic role of representing inertial effects only. By choosing an inertial frame, that connection can be made to vanish globally. Of course, as a purely inertial connection, its curvature vanishes identically:

$$\dot{R}^a{}_{b\mu\nu} = \partial_\mu \dot{A}^a{}_{b\nu} - \partial_\nu \dot{A}^a{}_{b\mu} + \dot{A}^a{}_{e\mu} \dot{A}^e{}_{b\nu} - \dot{A}^a{}_{e\nu} \dot{A}^e{}_{b\mu} = 0. \tag{4.60}$$

However, for a tetrad involving a non-trivial translational gauge potential $B^a{}_\mu$, that is, for

$$B^a{}_\mu \neq \dot{\mathscr{D}}_\mu \varepsilon^a, \tag{4.61}$$

torsion will be non-vanishing:

$$\dot{T}^a{}_{\mu\nu} = \partial_\mu h^a{}_\nu - \partial_\nu h^a{}_\mu + \dot{A}^a{}_{e\mu} h^e{}_\nu - \dot{A}^a{}_{e\nu} h^e{}_\mu \neq 0. \tag{4.62}$$

In Teleparallel Gravity, therefore, gravitation is represented by torsion, not by curvature. This is at variance with General Relativity, whose spin connection $\mathring{A}^a{}_{b\mu}$ has vanishing torsion

$$\mathring{T}^a{}_{\mu\nu} = \partial_\mu h^a{}_\nu - \partial_\nu h^a{}_\mu + \mathring{A}^a{}_{e\mu} h^e{}_\nu - \mathring{A}^a{}_{e\nu} h^e{}_\mu = 0, \tag{4.63}$$

but non-vanishing curvature

$$\mathring{R}^a{}_{b\mu\nu} = \partial_\mu \mathring{A}^a{}_{b\nu} - \partial_\nu \mathring{A}^a{}_{b\mu} + \mathring{A}^a{}_{e\mu} \mathring{A}^e{}_{b\nu} - \mathring{A}^a{}_{e\nu} \mathring{A}^e{}_{b\mu} \neq 0. \tag{4.64}$$

The spacetime-indexed linear connection corresponding to the inertial spin connection (4.46) is

$$\dot{\Gamma}^\rho{}_{\nu\mu} = h_a{}^\rho \partial_\mu h^a{}_\nu + h_a{}^\rho \dot{A}^a{}_{b\mu} h^b{}_\nu \equiv h_a{}^\rho \dot{\mathscr{D}}_\mu h^a{}_\nu. \tag{4.65}$$

This is the so-called Weitzenböck connection. Its definition is equivalent to the identity

$$\partial_\mu h^a{}_\nu + \dot{A}^a{}_{b\mu} h^b{}_\nu - \dot{\Gamma}^\rho{}_{\nu\mu} h^a{}_\rho = 0. \tag{4.66}$$

In the class of frames in which the spin connection $\dot{A}^a{}_{b\mu}$ vanishes, it reduces to

$$\partial_\mu h^a{}_\nu - \dot{\Gamma}^\rho{}_{\nu\mu} h^a{}_\rho = 0, \tag{4.67}$$

which is the absolute, or distant parallelism condition, from where Teleparallel Gravity got its name.

Comment 4.7 At the time Teleparallel Gravity was christened, no one was aware of the local Lorentz invariance of the theory. For this reason, the absolute distant condition (4.67) was believed to hold everywhere; hence the name Teleparallel Gravity. We know today, however, that this condition holds only in a very special class of Lorentz frames: that in which the teleparallel spin connection vanishes. The general, covariant condition is that given by Eq. (4.66), from where we see that the tetrad is not parallel-transported everywhere by the Weitzenböck connection. This means that the name Teleparallel Gravity is not appropriate. Of course, for historical reasons, we shall keep it.

The Weitzenböck connection $\dot{\Gamma}^{\rho}{}_{\mu\nu}$ is related to the Levi-Civita connection $\overset{\circ}{\Gamma}{}^{\rho}{}_{\mu\nu}$ of General Relativity by

$$\dot{\Gamma}^{\rho}{}_{\mu\nu} = \overset{\circ}{\Gamma}{}^{\rho}{}_{\mu\nu} + \dot{K}^{\rho}{}_{\mu\nu}, \tag{4.68}$$

where

$$\dot{K}^{\rho}{}_{\mu\nu} = \tfrac{1}{2}\left(\dot{T}_{\mu}{}^{\rho}{}_{\nu} + \dot{T}_{\nu}{}^{\rho}{}_{\mu} - \dot{T}^{\rho}{}_{\mu\nu}\right) \tag{4.69}$$

is the contortion of the Weitzenböck torsion

$$\dot{T}^{\rho}{}_{\nu\mu} = \dot{\Gamma}^{\rho}{}_{\mu\nu} - \dot{\Gamma}^{\rho}{}_{\nu\mu}. \tag{4.70}$$

It is important to observe that, whereas the spin connection (4.46) depends uniquely on the Lorentz transformations, and consequently only represents inertial effects, the corresponding spacetime-indexed connection (4.65) does depend on the gravitational field because it is obtained from (4.46) by contractions with tetrads, which depend on the gauge potential $B^{a}{}_{\mu}$. In the case of trivial (non-gravitational) tetrads $e^{a}{}_{\mu}$, instead of the Weitzenböck connection (4.65), one obtains the inertial connection

$$\dot{\gamma}^{\rho}{}_{\nu\mu} = e_{a}{}^{\rho}\partial_{\mu}e^{a}{}_{\nu} + e_{a}{}^{\rho}\dot{A}^{a}{}_{b\mu}e^{b}{}_{\nu}. \tag{4.71}$$

This is the connection (2.21), which appears in the equation of motion of free particles when described in a general coordinate system. In cartesian coordinates, and considering the class of frames in which the teleparallel spin connection vanishes, this connection vanishes as well.

Comment 4.8 It should be remarked that R. Weitzenböck, although has worked with torsion gravity [20], does not seem to have ever written connection (4.65). Though not supported by historical fact, the name "Weitzenböck connection" is commonly used to denote this particular case of a Cartan connection, a practice we shall bow to.

References

1. Weyl, H.: Gravitation und Elektrizität. Sitzungsber. Preuss. Akad. Wissensch., p. 465 (1918)
2. O'Raifeartaigh, L.: The Dawning of Gauge Theory. Princeton University Press, Princeton (1998)
3. O'Raifeartaigh, L., Straumann, N.: Rev. Mod. Phys. **72**, 1 (2000). hep-ph/9810524
4. Einstein, A.: Auf die Riemann-Metrik und den Fern-Parallelismus gegründete einheitliche Feldtheorie. Math. Ann. **102**, 685 (1930). For an English translation, together with three previous essays, see Unzicker, A., Case, T.: Unified field theory based on Riemannian metrics and distant parallelism. physics/0503046
5. Sauer, T.: Field equations in teleparallel spacetime: Einstein's 'Fernparallelismus' approach towards unified field theory. Einstein's Papers Project. physics/0405142
6. Schücking, E.L.: Einstein's apple and relativity's gravitational field. 0903.3768 [physics]
7. Kaluza, Th.: Sitzungsber. Preuss. Akad. Wissensch. Phys. Math. Kl., p. 966 (1921)
8. Klein, O.: Z. Phys. **37**, 895 (1926)
9. Cartan, É.: Ann. Sci. Éc. Norm. Super. **40**, 325 (1923)
10. Cartan, É.: Ann. Sci. Éc. Norm. Super. **41**, 1 (1924)
11. Møller, C.: K. Dan. Vidensk. Selsk. Mat. Fys. Skr. **1**, 10 (1961)

12. Pellegrini, C., Plebanski, J.: K. Dan. Vidensk. Selsk. Mat. Fys. Skr. **2**, 2 (1962)
13. Møller, C.: K. Dan. Vidensk. Selsk. Mat. Fys. Skr. **89**, 13 (1978)
14. Hayashi, K., Nakano, T.: Prog. Theor. Phys. **38**, 491 (1967)
15. Hayashi, K.: Nuovo Cimento A **16**, 639 (1973)
16. Cho, Y.M.: Phys. Rev. D **14**, 2521 (1976)
17. Hayashi, K.: Phys. Lett. B **69**, 441 (1977)
18. Hayashi, K., Shirafuji, T.: Phys. Rev. D **19**, 3524 (1979)
19. Aldrovandi, R., Pereira, J.G.: An Introduction to Geometrical Physics. World Scientific, Singapore (1995)
20. Weitzenböck, R.: Invarianten-Theorie. Noordhoff, Groningen (1923)

Chapter 5
Gravitational Coupling Prescription

The gravitational coupling prescription is made up of two parts: a translational coupling prescription, which is universal and comes from the translational gauge invariance, and a Lorentz coupling prescription, which is non-universal and comes from the requirement of local Lorentz invariance. Together, they yield the full coupling prescription of any field to gravitation. The specific cases of General Relativity and Teleparallel Gravity are discussed, and their equivalence established.

5.1 Translational Coupling Revisited

As we have seen in the previous chapter, the covariance under local translations required the introduction of a gauge potential, which gave rise to a translational covariant derivative. This covariant derivative defined a translational coupling prescription, according to which a trivial tetrad was replaced by a non-trivial one related to a gravitational field:

$$e^a{}_\mu \to h^a{}_\mu. \tag{5.1}$$

Since these tetrads satisfy the relations

$$\eta_{\mu\nu} = e^a{}_\mu e^b{}_\nu \eta_{ab} \quad \text{and} \quad g_{\mu\nu} = h^a{}_\mu h^b{}_\nu \eta_{ab}, \tag{5.2}$$

with $\eta_{\mu\nu}$ the Minkowski and $g_{\mu\nu}$ a general riemannian metric related to a gravitational field, the translational coupling prescription is ultimately equivalent to the metric replacement

$$\eta_{\mu\nu} \to g_{\mu\nu}. \tag{5.3}$$

5.2 Lorentz Coupling Prescription

In addition to being invariant under local translations, any theory must also be invariant under local Lorentz transformations. This second invariance is related to the fact

R. Aldrovandi, J.G. Pereira, *Teleparallel Gravity*, Fundamental Theories of Physics 173, DOI 10.1007/978-94-007-5143-9_5,

that physics must be the same, independently of the frame used to describe it. Under a local Lorentz transformation one changes from a given class of frames to another one, which differs from the first by the presence of different inertial effects. Although not a dynamic (or gauge) symmetry, local Lorentz invariance introduces an additional coupling prescription, which also amounts to replace all ordinary derivatives by a Lorentz covariant derivative,

$$\partial_\mu \to \mathscr{D}_\mu = \partial_\mu + \tfrac{i}{2} A^{ab}{}_\mu S_{ab}, \tag{5.4}$$

with S_{ab} the Lorentz generators appropriate for the field under consideration. Of course, differently from the translation coupling prescription (5.1), the Lorentz part of the coupling is not universal, in the sense that it depends on the spin content of each field (or particle), that is, on the Lorentz representation to which it belongs.

5.2.1 General Covariance Principle

The Lorentz part of the gravitational coupling prescription can be obtained from the so-called *general covariance principle* [1]. This principle states that an equation valid in Special Relativity can be made to hold in the presence of gravitation if it is written in a generally covariant form, that is, if it preserves its form under general coordinate transformations. In order to make an equation generally covariant, it is necessary to introduce a connection which is in principle concerned only with the *inertial* properties of the coordinate system under consideration. As long as only coordinate transformations are involved, just a vacuum connection (with zero curvature and torsion) is needed. Then, by using the local equivalence between inertial and gravitational effects, that connection can be replaced by a connection representing a *true gravitational field*. In this way, equations valid in the presence of gravitation are obtained from the corresponding equations holding in Special Relativity.

The general covariance principle can be seen as an *active* version of the strong equivalence principle in the sense that, by making a special-relativistic equation covariant and using the strong equivalence principle, it is possible to obtain its form in the presence of gravitation. The usual form of the equivalence principle, on the other hand, can be interpreted as its *passive* version: given an equation valid in the presence of gravitation, the corresponding special-relativistic equation must be recovered locally, or more precisely, at a point or at most along a trajectory [2]. It should be emphasized that general covariance by itself is empty of physical content, as any equation can be *made* generally covariant. Only when use is made of the local equivalence between inertial and gravitational effects, and the compensating term is replaced with a gravitational connection, can the principle of general covariance be seen as an active version of the strong equivalence principle [3].

The above description of the general covariance principle refers to its usual *holonomic* version, based on coordinate transformations. An alternative, more general version of the principle can be obtained by using anholonomic frames, based on local Lorentz transformations. The basic difference between these two versions is that, instead of requiring that an equation be covariant under general coordinate transformations, in the frame version the equation is required to be covariant under local Lorentz transformations. In spite of the different nature of the involved transformations, the physical content of both approaches is the same (see Comment 5.1). The frame version is, however, more general: unlike the coordinate version, which can be used for tensor fields only, the frame version holds for both integer and half-integer spin fields [4].

An important point of the general covariance principle is that it defines in a natural way a Lorentz-covariant derivative, and consequently also a gravitational coupling prescription. The process of obtaining this coupling prescription comprises then two steps. The first is to pass to a general anholonomic frame, where inertial effects—which appear in the form of a compensating term, or vacuum Lorentz connection—are present. Then, by using the strong equivalence principle, the compensating term is replaced by a connection representing, instead of inertial effects, a gravitational field, yielding in this way a gravitational coupling prescription.

Comment 5.1 Equation (2.23) says that the inertial connection (2.7), obtained by performing a *local Lorentz transformation*, is equivalent to the spacetime-indexed inertial connection (2.21), which is obtained by performing a *general coordinate transformation*. Namely, they represent two different ways of expressing the very same inertial connection. This means that local Lorentz and general coordinate transformation (or diffeomorphism) are equivalent transformations. Considering that diffeomorphism is empty of dynamical meaning, local Lorentz transformation is consequently also empty of dynamical meaning. This is on the root of the general covariance principle discussed above.

5.2.2 Passage to an Anholonomic Frame

The first step to obtain the Lorentz coupling prescription is to move to a general anholonomic frame. Let us then consider a vector field ϕ'^c on Minkowski spacetime. In the trivial, holonomic frame [see Comment 4.4]

$$e'_a = \delta_a^\mu \partial_\mu \tag{5.5}$$

with components δ_a^μ, its ordinary derivative is

$$e'_a \phi'^c = \delta_a^\mu \partial_\mu \phi'^c. \tag{5.6}$$

Under a local Lorentz transformation $\Lambda^d{}_c(x)$, a vector field transforms according to

$$\phi^c = \Lambda_d{}^c(x) \phi'^d. \tag{5.7}$$

The original and the Lorentz-transformed derivatives are then related by

$$e'_a \phi'^c = \Lambda^b{}_a(x) \Lambda_d{}^c(x) \mathscr{D}_b \phi^d, \tag{5.8}$$

where

$$\mathscr{D}_b \phi^d = h_b \phi^d + \Lambda^d{}_e(x) e_b \Lambda_c{}^e(x) \phi^c, \tag{5.9}$$

and

$$e_b = \Lambda_b{}^a(x) e'_a \tag{5.10}$$

is the transformed frame, which is of course anholonomic:

$$[e_b, e_c] = f^a{}_{bc} e_a \neq 0. \tag{5.11}$$

Making use of the orthogonality property of the tetrads, we see from Eq. (5.10) that the element of the Lorentz group can be written in the form

$$\Lambda_b{}^d(x) = e_b{}^\rho \delta^d_\rho. \tag{5.12}$$

From this expression, and using the commutation relation (5.11), the connection term appearing in the covariant derivative (5.9) can be rewritten in the form

$$\Lambda^c{}_d(x) e_a \Lambda_b{}^d(x) = \tfrac{1}{2}\big(f_b{}^c{}_a + f_a{}^c{}_b - f^c{}_{ba}\big). \tag{5.13}$$

Substituting in the covariant derivative (5.9), it becomes

$$\mathscr{D}_a \phi^c = h_a \phi^c + \tfrac{1}{2}\big(f_b{}^c{}_a + f_a{}^c{}_b - f^c{}_{ba}\big)\phi^b. \tag{5.14}$$

The freedom to choose any tetrad $\{e_a\}$ as a moving frame on Minkowski spacetime introduces the compensating term $\frac{1}{2}(f_b{}^c{}_a + f_a{}^c{}_b - f^c{}_{ba})$ in the derivative of a vector field. This term is, of course, concerned only with the inertial properties of that frame. In other words, it represents the inertial effects inherent to the chosen frame.

5.2.3 Identifying Inertia with Gravitation

Let us consider the relation (1.59), which is valid for a general Lorentz connection with arbitrary curvature and torsion. It can be rewritten in the form

$$A^c{}_{ba} - A^c{}_{ab} = T^c{}_{ab} + f^c{}_{ab}, \tag{5.15}$$

where $T^c{}_{ab}$, we recall, is the torsion tensor. Use of this equation for three different combination of indices gives

$$\tfrac{1}{2}\big(f_b{}^c{}_a + f_a{}^c{}_b - f^c{}_{ba}\big) = A^c{}_{ba} - K^c{}_{ba}, \tag{5.16}$$

where

$$K^c{}_{ba} = \tfrac{1}{2}\big(T_b{}^c{}_a + T_a{}^c{}_b - T^c{}_{ba}\big) \tag{5.17}$$

is the contortion tensor in the tetrad frame. Differently from the covariant derivative (5.14), where the coefficient of anholonomy $f^c{}_{ba}$ represents inertial effects in Minkowski spacetime, in expression (5.16) the coefficient of anholonomy $f^c{}_{ba}$ represents both gravitation and inertial effects. According to the general covariance principle, therefore, substituting (5.16) in the covariant derivative (5.14), which is a flat spacetime covariant derivative as seen in a general frame, we obtain the covariant derivative valid in the presence of gravitation:

$$\mathscr{D}_a \phi^c = h_a \phi^c + \big(A^c{}_{ba} - K^c{}_{ba}\big)\phi^b. \tag{5.18}$$

In terms of the vector representation

$$(S_{eb})^c{}_d = i\left(\delta_e^c \eta_{bd} - \delta_b^c \eta_{ed}\right) \tag{5.19}$$

of the Lorentz generators, it assumes the form

$$\mathscr{D}_a \phi^c = h_a \phi^c - \tfrac{i}{2}\left(A^{eb}{}_a - K^{eb}{}_a\right)(S_{eb})^c{}_d \phi^d. \tag{5.20}$$

Although obtained in the specific case of a Lorentz vector field, the compensating term (5.13) can be easily verified to be the same for any representation. In fact, considering a general source field Ψ carrying an arbitrary representation of the Lorentz group, its Lorentz transformation will be

$$\Psi' = U(\Lambda)\Psi, \tag{5.21}$$

where

$$U(\Lambda) = \exp\left(\tfrac{i}{2}\varepsilon^{bc} S_{bc}\right)$$

is the element of the Lorentz group in the arbitrary representation S_{bc}. As a simple calculation shows [4],

$$U(\Lambda) e_a U^{-1}(\Lambda) = \tfrac{i}{4}\left(f^{bc}{}_a + f_a{}^{cb} - f^{cb}{}_a\right) S_{bc}. \tag{5.22}$$

In this general case, the covariant derivative (5.20) assumes the form

$$\mathscr{D}_a \Psi = h_a \Psi - \tfrac{i}{2}\left(A^{bc}{}_a - K^{bc}{}_a\right) S_{bc} \Psi. \tag{5.23}$$

This is the covariant derivative defining the Lorentz part of the gravitational coupling prescription.

5.3 Full Gravitational Coupling Prescription

The full gravitational coupling prescription is then composed of two parts: one, corresponding to the (universal) translational coupling prescription, which is represented by

$$e^a{}_\mu \partial_a \Psi \to h^a{}_\mu \partial_a \Psi, \tag{5.24}$$

and another, corresponding to the (non-universal) Lorentz coupling prescription, represented by

$$\partial_a \Psi \to \mathscr{D}_a \Psi. \tag{5.25}$$

Put together, they yield the *full gravitational coupling prescription*,

$$e^a{}_\mu \partial_a \Psi \to h^a{}_\mu \mathscr{D}_a \Psi = h^a{}_\mu \left[h_a \Psi - \tfrac{i}{2}\left(A^{bc}{}_a - K^{bc}{}_a\right) S_{bc} \Psi\right]. \tag{5.26}$$

Equivalently, one can write

$$\partial_\mu \Psi \to \mathscr{D}_\mu \Psi = \partial_\mu \Psi - \tfrac{i}{2}\left(A^{ab}{}_\mu - K^{ab}{}_\mu\right) S_{ab} \Psi, \tag{5.27}$$

where now it is understood that, after the application of the coupling prescription, the spacetime indices $\mu, \nu, \rho \ldots$ are to be raised and lowered with the spacetime metric

$$g_{\mu\nu} = \eta_{ab} h^a{}_\mu h^b{}_\nu. \tag{5.28}$$

It is important to emphasize once more that this is the gravitational coupling prescription that follows from the general covariance principle, that is to say, from the strong equivalence principle. Any other form of the coupling prescription will be in contradiction with that principle.

5.4 Possible Connections

A crucial point of the general covariance principle is that it does not determine uniquely the Lorentz connection $A^{bc}{}_\mu$. In fact, from the point of view of the coupling prescription, the connection can be chosen freely among the infinitely many possibilities, each one characterized by a connection with different values of curvature and torsion. However, due to the identity

$$A^{bc}{}_\mu - K^{bc}{}_\mu = \overset{\circ}{A}{}^{bc}{}_\mu, \tag{5.29}$$

with $\overset{\circ}{A}{}^{bc}{}_\mu$ the (torsionless) spin connection of General Relativity, any one of the choices will give rise to a coupling prescription that is ultimately equivalent to the coupling prescription of General Relativity:

$$\partial_\mu \Psi \to \overset{\circ}{\mathscr{D}}_\mu \Psi = \partial_\mu \Psi - \tfrac{i}{2} \overset{\circ}{A}{}^{bc}{}_\mu S_{bc} \Psi. \tag{5.30}$$

According to the general covariance principle, therefore, *the gravitational coupling prescription is "minimal" only with the spin connection of General Relativity*. Among other consequences, we see that only in General Relativity the equation of motion of spinless particles can be described by auto-parallel curves, or geodesics [see Chap. 6 for further details].

Comment 5.2 Although a general Lorentz connection $A^{bc}{}_\mu$ (with non-vanishing curvature and torsion) has twenty four independent components, on account of the identity (5.29), the difference $A^{bc}{}_\mu - K^{bc}{}_\mu$ is seen to be equivalent to the General Relativity spin connection, which is fully determined by the metric (or by the tetrad). This means that the coupling prescription (5.27) does not introduce any additional degree of freedom in relation to General Relativity. Although it involves a connection presenting curvature and torsion, therefore, it is consistent with any gravitational model in which the source is the ten-components symmetric energy-momentum tensor. As a complementary discussion to this point, in Appendix B the properties of the space of Lorentz connections are studied in some more detail.

What has been said holds, in particular, for the purely inertial connection

$$\overset{\bullet}{A}{}^b{}_{c\mu} = \Lambda^b{}_d(x) \partial_\mu \Lambda_c{}^d(x) \tag{5.31}$$

in the sense that in this case the identity reads

$$\overset{\bullet}{A}{}^{bc}{}_\mu - \overset{\bullet}{K}{}^{bc}{}_\mu = \overset{\circ}{A}{}^{bc}{}_\mu. \tag{5.32}$$

As we have seen in Chap. 4, the gravitational theory corresponding to this choice is just Teleparallel Gravity. In this theory, the gravitational field is fully represented by the gauge potential $B^a{}_\mu$, which appears as the non-trivial part of the tetrad field. In Teleparallel Gravity, therefore, the gravitational coupling prescription that follows from the general covariance principle is [5]

$$\partial_\mu \Psi \to \overset{\bullet\bullet}{\mathscr{D}}_\mu \Psi = \partial_\mu \Psi - \tfrac{i}{2}\big(\overset{\bullet}{A}{}^{bc}{}_\mu - \overset{\bullet}{K}{}^{bc}{}_\mu\big) S_{bc} \Psi, \tag{5.33}$$

with $\overset{\bullet}{K}{}^{bc}{}_\mu$ the contortion of the connection $\overset{\bullet}{A}{}^{bc}{}_\mu$. In the specific case of a Lorentz vector field ϕ^b, for which S_{bc} is given by Eq. (5.19), the teleparallel coupling prescription assumes the form

$$\partial_\mu \phi^b \to \overset{\bullet\bullet}{\mathscr{D}}_\mu \phi^b = \partial_\mu \phi^b + \big(\overset{\bullet}{A}{}^b{}_{c\mu} - \overset{\bullet}{K}{}^b{}_{c\mu}\big)\phi^c. \tag{5.34}$$

The corresponding expression for the spacetime vector $\phi^\rho = h_c{}^\rho \phi^c$ is

$$\partial_\mu \phi^\rho \to \overset{\bullet\bullet}{\nabla}_\mu \phi^\rho = \partial_\mu \phi^\rho + \big(\overset{\bullet}{\Gamma}{}^\rho{}_{\lambda\mu} - \overset{\bullet}{K}{}^\rho{}_{\lambda\mu}\big)\phi^\lambda. \tag{5.35}$$

These two derivatives are easily seen to be related by

$$\overset{\bullet\bullet}{\mathscr{D}}_\mu \phi^b = h^b{}_\rho \overset{\bullet\bullet}{\nabla}_\mu \phi^\rho.$$

Of course, due to the identity

$$\overset{\bullet}{A}{}^{bc}{}_\mu - \overset{\bullet}{K}{}^{bc}{}_\mu = \overset{\circ}{A}{}^{bc}{}_\mu, \tag{5.36}$$

the above coupling prescription is also equivalent to the coupling prescription of General Relativity:

$$\overset{\bullet\bullet}{\mathscr{D}}_\mu \Psi = \overset{\circ}{\mathscr{D}}_\mu \Psi. \tag{5.37}$$

However, although physically equivalent, the coupling prescription (5.33) is, conceptually speaking, completely different. In particular, since the spin connection (5.31) represents inertial effects only, the gravitational interaction turns out to be described in a completely different way. In Chap. 6, where the motion of a test particle will be studied in the context of both General Relativity and Teleparallel Gravity, a thorough analysis of these differences will be presented. The same analysis, but for the known fundamental fields of Nature, will be performed in Chaps. 12 and 13.

References

1. Weinberg, S.: Gravitation and Cosmology. Wiley, New York (1972)
2. Aldrovandi, R., Barros, P.B., Pereira, J.G.: Found. Phys. **33**, 545 (2003). gr-qc/0212034
3. Sciama, D.W.: Rev. Mod. Phys. **36**, 463 (1964)
4. Mosna, R.A., Pereira, J.G.: Gen. Relativ. Gravit. **36**, 2525 (2004). gr-qc/0312093
5. de Andrade, V.C., Guillen, L.C.T., Pereira, J.G.: Phys. Rev. D **64**, 027502 (2001). gr-qc/0104102

Chapter 6
Particle Mechanics

The teleparallel force equation is obtained from a variational principle, its equivalence with the geodesic equation is established, and the roles played by torsion and curvature in the descriptions of the gravitational interaction are discussed. The newtonian limit is studied, and the gravitomagnetic field is obtained in terms of torsion components. Because the teleparallel spin connection represents inertial effects only, Teleparallel Gravity produces a separation between inertial effects and gravitation. The gravitational field in this theory is represented by the translational gauge potential, a true field variable in the usual sense of field theory and a genuine gravitational connection.

6.1 Free Particles Revisited

We will here re-obtain the equation of motion of a free particle studied in Sect. 2.2, but now using a variational principle. This is intended as a warm up exercise, whose scheme will be used later to obtain the gravitationally-coupled equation of motion in Teleparallel Gravity.

6.1.1 Basic Notions

Let us start with Minkowski spacetime, whose invariant quadratic interval is

$$d\sigma^2 = \eta_{\mu\nu}\, dx^\mu\, dx^\nu, \tag{6.1}$$

where $\eta_{\mu\nu} = \eta_{ab} e^a{}_\mu e^b{}_\nu$ is the Minkowski spacetime metric, with $e^a{}_\mu$ a trivial, inertial tetrad. Its most general form is

$$e^a{}_\mu \equiv \overset{\bullet}{\mathscr{D}}_\mu x^a = \partial_\mu x^a + \overset{\bullet}{A}{}^a{}_{b\mu} x^b, \tag{6.2}$$

with $\overset{\bullet}{A}{}^a{}_{b\mu}$ the inertial connection (5.31). In terms of this tetrad, the quadratic interval (6.1) reads

$$d\sigma^2 = \eta_{ab} e^a e^b, \tag{6.3}$$

R. Aldrovandi, J.G. Pereira, *Teleparallel Gravity*, Fundamental Theories of Physics 173, DOI 10.1007/978-94-007-5143-9_6, © Springer Science+Business Media Dordrecht 2013

with $e^a = e^a{}_\mu \, dx^\mu$ given by

$$e^a = dx^a + \overset{\bullet}{A}{}^a{}_b x^b. \tag{6.4}$$

The holonomic particle four-velocity is

$$u^\mu = \frac{dx^\mu}{d\sigma}. \tag{6.5}$$

Hence, along the trajectory of the particle, the spacetime interval can be written in the form

$$d\sigma = u_\mu \, dx^\mu. \tag{6.6}$$

In terms of the anholonomic four-velocity

$$u^a \equiv e^a{}_\mu u^\mu = e^a\left(\frac{d}{d\sigma}\right), \tag{6.7}$$

the same spacetime interval is written as

$$d\sigma = u_a e^a. \tag{6.8}$$

6.1.2 Equation of Motion of Free Particles

A free particle of mass m is represented by the action integral

$$\mathcal{S} = -mc \int_{\mathrm{p}}^{\mathrm{q}} d\sigma = -mc \int_{\mathrm{p}}^{\mathrm{q}} u_a e^a, \tag{6.9}$$

where we have used identity (6.8). Substituting Eq. (6.4), it assumes the form

$$\mathcal{S} = -mc \int_{\mathrm{p}}^{\mathrm{q}} u_a \big(dx^a + \overset{\bullet}{A}{}^a{}_{b\mu} x^b \, dx^\mu\big). \tag{6.10}$$

Under a general spacetime variation $x^\mu \to x^\mu + \delta x^\mu$, it changes according to

$$\delta\mathcal{S} = -mc \int_{\mathrm{p}}^{\mathrm{q}} \big[e^a \delta u_a + u_a d\delta x^a + u_a \delta\big(\overset{\bullet}{A}{}^a{}_{b\mu} x^b\big) dx^\mu + u_a \overset{\bullet}{A}{}^a{}_{b\mu} x^b d\big(\delta x^\mu\big)\big], \tag{6.11}$$

where we have already used the property $[\delta, d] = 0$.

Comment 6.1 Writing the quadratic spacetime interval in the form

$$d\sigma^2 = \eta_{ab} e^a e^b, \tag{6.12}$$

a direct calculation shows that

$$\delta(d\sigma) = u_a \delta e^a. \tag{6.13}$$

On the other hand, from Eq. (6.8), we get

$$\delta(d\sigma) = u_a \delta e^a + e^a \delta u_a. \tag{6.14}$$

Comparing Eqs. (6.13) and (6.14), we see immediately that

$$e^a \delta u_a = 0. \tag{6.15}$$

Taking into account Eq. (6.15), the action variation (6.11) becomes

$$\delta\mathcal{S} = -mc\int_{\mathsf{p}}^{\mathsf{q}}\left[u_a\,d\delta x^a + u_a\delta\left(\overset{\bullet}{A}{}^a{}_{b\mu}x^b\right)dx^\mu + u_a\overset{\bullet}{A}{}^a{}_{b\mu}x^b\,d\left(\delta x^\mu\right)\right]. \quad (6.16)$$

Integrating by parts the first and the third terms and neglecting the surface terms, we obtain

$$\delta\mathcal{S} = mc\int_{\mathsf{p}}^{\mathsf{q}}\left[du_a\delta x^a - u_a\delta\left(\overset{\bullet}{A}{}^a{}_{b\mu}x^b\right)dx^\mu + d\left(u_a\overset{\bullet}{A}{}^a{}_{b\mu}x^b\right)\delta x^\mu\right]. \quad (6.17)$$

Performing the variations and differentials, and using the expressions

$$\delta x^a = \partial_\mu x^a\delta x^\mu \quad \text{and} \quad \delta\overset{\bullet}{A}{}^a{}_{b\mu} = \partial_\rho\overset{\bullet}{A}{}^a{}_{b\mu}\delta x^\rho$$

we get, after a straightforward algebra,

$$\delta\mathcal{S} = mc\int_{\mathsf{p}}^{\mathsf{q}}\left[e^a{}_\mu\left(\frac{du_a}{d\sigma} - \overset{\bullet}{A}{}^b{}_{a\rho}u_b u^\rho\right) - \overset{\bullet}{R}{}^a{}_{b\mu\rho}x^b u_a u^\rho\right]d\sigma\,\delta x^\mu. \quad (6.18)$$

Since the curvature $\overset{\bullet}{R}{}^a{}_{b\mu\rho}$ of the inertial connection $\overset{\bullet}{A}{}^a{}_{b\mu}$ vanishes identically, we are actually left with

$$\delta\mathcal{S} = mc\int_{\mathsf{p}}^{\mathsf{q}}\left[e^a{}_\mu\left(\frac{du_a}{d\sigma} - \overset{\bullet}{A}{}^b{}_{a\rho}u_b u^\rho\right)\right]d\sigma\,\delta x^\mu. \quad (6.19)$$

The invariance of the action, together with the arbitrariness of δx^μ, yields then the equation of motion

$$u^\rho\overset{\bullet}{\mathcal{D}}_\rho u_a \equiv \frac{du_a}{d\sigma} - \overset{\bullet}{A}{}^b{}_{a\rho}u_b u^\rho = 0, \quad (6.20)$$

whose contravariant form is

$$u^\rho\overset{\bullet}{\mathcal{D}}_\rho u^a \equiv \frac{du^a}{d\sigma} + \overset{\bullet}{A}{}^a{}_{b\rho}u^b u^\rho = 0. \quad (6.21)$$

This is the equation of motion of a free particle as seen from the general Lorentz frame e^a. Of course, in the class of holonomic frames e'^a, the inertial connection $\overset{\bullet}{A}{}'^a{}_{b\rho}$ vanishes and the equation of motion reduces to

$$\frac{du'^a}{d\sigma} = 0. \quad (6.22)$$

6.2 Gravitationally Coupled Particles

We are going now to use the same procedure of the last section to obtain the equation of motion of a test particle in the presence of gravitation.

6.2.1 Coupling Prescription

The gravitational coupling prescription is carried out by replacing a trivial tetrad representing Minkowski spacetime by a nontrivial tetrad representing a gravitational field:

$$e^a{}_\mu \to h^a{}_\mu. \tag{6.23}$$

In the specific case of Teleparallel Gravity, this is achieved by replacing

$$e^a{}_\mu = \overset{\bullet}{\mathscr{D}}_\mu x^a \quad \to \quad h^a{}_\mu = \overset{\bullet}{\mathscr{D}}_\mu x^a + B^a{}_\mu, \tag{6.24}$$

with $B^a{}_\mu$ the translational gauge potential. In consonance with this replacement, the spacetime metric changes according to

$$\eta_{\mu\nu} = \eta_{ab} e^a{}_\mu e^b{}_\nu \quad \to \quad g_{\mu\nu} = \eta_{ab} h^a{}_\mu h^b{}_\nu, \tag{6.25}$$

which implies the further change

$$d\sigma^2 = \eta_{\mu\nu}\, dx^\mu\, dx^\nu \quad \to \quad ds^2 = g_{\mu\nu}\, dx^\mu\, dx^\nu. \tag{6.26}$$

In terms of anholonomic bases, this last expression assumes the form

$$d\sigma^2 = \eta_{ab} e^a e^b \quad \to \quad ds^2 = \eta_{ab} h^a h^b. \tag{6.27}$$

Along the particle trajectory, therefore, the spacetime interval can be written formally as

$$ds = g_{\mu\rho} u^\mu\, dx^\rho = \eta_{ab} u^a h^b, \tag{6.28}$$

with

$$u^\mu = \frac{dx^\mu}{ds} \tag{6.29}$$

the holonomic four-velocity of the particle, which is related to the anholonomic four-velocity u^a by

$$u^a \equiv h^a{}_\mu u^\mu = h^a \left(\frac{d}{ds}\right). \tag{6.30}$$

6.2.2 Gravitationally Coupled Equation of Motion

The action representing a particle of mass m in the presence of gravitation is

$$\mathscr{S} = -mc \int_{\mathrm{p}}^{\mathrm{q}} ds = -mc \int_{\mathrm{p}}^{\mathrm{q}} u_a h^a, \tag{6.31}$$

where we have used Eq. (6.28). Substituting

$$h^a = dx^a + \overset{\bullet}{A}{}^a{}_{b\mu} x^b\, dx^\mu + B^a{}_\mu\, dx^\mu, \tag{6.32}$$

it becomes

$$\mathscr{S} = -mc \int_{\mathsf{p}}^{\mathsf{q}} u_a \big(dx^a + \overset{\bullet}{A}{}^a{}_{b\mu} x^b \, dx^\mu + B^a{}_\mu \, dx^\mu \big). \tag{6.33}$$

This is the teleparallel version of the action, as described from a general Lorentz frame. The first term in the integrand represents the free particle, the second represents the interaction with the inertial effects, and the third represents the gravitational interaction. In the class of frames e'^a in which the inertial connection $\overset{\bullet}{A}{}'^a{}_b$ vanishes, it reduces to

$$\mathscr{S} = -mc \int_{\mathsf{p}}^{\mathsf{q}} u'_a \big(dx'^a + B'^a{}_\mu \, dx^\mu \big). \tag{6.34}$$

Comment 6.2 Observe that, due to the gauge structure of Teleparallel Gravity, the action has a form similar to the action of a charged particle in an electromagnetic field. In fact, if the particle has additionally an electric charge q and is in the presence of an electromagnetic potential A_μ, the total action has the form

$$\mathscr{S} = -mc \int_{\mathsf{p}}^{\mathsf{q}} \left[u_a \, dx^a + u_a \overset{\bullet}{A}{}^a{}_{b\mu} x^b \, dx^\mu + u_a B^a{}_\mu \, dx^\mu + \frac{q}{mc^2} A_\mu \, dx^\mu \right]. \tag{6.35}$$

Notice that, whereas the electromagnetic interaction depends on the relation q/m of the particle, the gravitational interaction has already been assumed to be universal in the sense that it does not depend on any specific property of the particle. In Chap. 11 we will present a more detailed discussion of this point.

Under a general spacetime variation $x^\mu \to x^\mu + \delta x^\mu$, action (6.33) changes according to

$$\delta\mathscr{S} = -mc \int_{\mathsf{p}}^{\mathsf{q}} \big(h^a \delta u_a + u_a \, d\delta x^a + u^a \delta \overset{\bullet}{A}{}^a{}_{b\mu} x^b \, dx^\mu + u_a \overset{\bullet}{A}{}^a{}_{b\mu} \delta x^b \, dx^\mu$$
$$+ u^a \overset{\bullet}{A}{}^a{}_{b\mu} x^b \, d\delta x^\mu + u_a \delta B^a{}_\mu \, dx^\mu + u_a B^a{}_\mu \, d\delta x^\mu \big), \tag{6.36}$$

where we have already used that $[\delta, d] = 0$.

Comment 6.3 Writing the quadratic spacetime interval in the form

$$ds^2 = \eta_{ab} h^a h^b, \tag{6.37}$$

a direct calculation shows that

$$\delta(ds) = u_a \delta h^a. \tag{6.38}$$

On the other hand, writing the interval in the form

$$ds = u_a h^a, \tag{6.39}$$

we get

$$\delta(ds) = u_a \delta h^a + h^a \delta u_a. \tag{6.40}$$

Comparing Eqs. (6.38) and (6.40), we see immediately that

$$h^a \delta u_a = 0. \tag{6.41}$$

Substituting (6.41) in the action variation (6.36), it becomes

$$\delta\mathcal{S} = -mc \int_{\mathsf{p}}^{\mathsf{q}} \big(u_a\, d\delta x^a + u_a \delta \dot{A}^a{}_{b\mu} x^b\, dx^\mu + u_a \dot{A}^a{}_{b\mu} \delta x^b\, dx^\mu$$
$$+ u_a \dot{A}^a{}_{b\mu} x^b\, d\delta x^\mu + u_a \delta B^a{}_\mu\, dx^\mu + u_a B^a{}_\mu\, d\delta x^\mu\big). \tag{6.42}$$

Integrating by parts the terms containing differentials of variations and neglecting the surface terms, we get

$$\delta\mathcal{S} = mc \int_{\mathsf{p}}^{\mathsf{q}} \big[du_a \delta x^a - u_a \delta \dot{A}^a{}_{b\mu} x^b\, dx^\mu - u_a \dot{A}^a{}_{b\mu} \delta x^b\, dx^\mu$$
$$+ d\big(u_a \dot{A}^a{}_{b\mu} x^b\big)\delta x^\mu - u_a \delta B^a{}_\mu dx^\mu + d\big(u_a B^a{}_\mu\big)\delta x^\mu\big]. \tag{6.43}$$

Performing the differentials and variations, substituting the expressions

$$\delta x^a = \partial_\mu x^a \delta x^\mu, \qquad \delta \dot{A}^a{}_{b\mu} = \partial_\rho \dot{A}^a{}_{b\mu} \delta x^\rho, \qquad \delta B^a{}_\mu = \partial_\rho B^a{}_\mu \delta x^\rho, \tag{6.44}$$

and considering that the curvature of the teleparallel spin connection $\dot{A}^a{}_{b\rho}$ vanishes, we get finally

$$\delta\mathcal{S} = mc \int_{\mathsf{p}}^{\mathsf{q}} \left[h^a{}_\mu \left(\frac{du_a}{ds} - \dot{A}^b{}_{a\rho} u_b u^\rho\right) - \dot{T}^b{}_{\mu\rho} u_b u^\rho\right] \delta x^\mu\, ds, \tag{6.45}$$

where

$$\dot{T}^a{}_{\mu\rho} = \dot{\mathcal{D}}_\mu B^a{}_\rho - \dot{\mathcal{D}}_\rho B^a{}_\mu \tag{6.46}$$

is the translational field strength, or torsion. From the invariance of the action, and taking into account the arbitrariness of δx^μ, the equation of motion is found to be

$$\frac{du_a}{ds} - \dot{A}^b{}_{a\rho} u_b u^\rho = \dot{T}^b{}_{a\rho} u_b u^\rho. \tag{6.47}$$

Its contravariant version is

$$\frac{du^a}{ds} + \dot{A}^a{}_{b\rho} u^b u^\rho = \dot{T}_b{}^a{}_\rho u^b u^\rho. \tag{6.48}$$

Using the identity

$$\dot{T}^b{}_{a\rho} u_b u^\rho = -\dot{K}^b{}_{a\rho} u_b u^\rho, \tag{6.49}$$

which follows from the contortion definition (1.63), they can be rewritten, respectively, in the forms

$$\frac{du_a}{ds} - \dot{A}^b{}_{a\rho} u_b u^\rho = -\dot{K}^b{}_{a\rho} u_b u^\rho \tag{6.50}$$

and

$$\frac{du^a}{ds} + \dot{A}^a{}_{b\rho} u^b u^\rho = \dot{K}^a{}_{b\rho} u^b u^\rho. \tag{6.51}$$

These are the teleparallel equations of motion of a particle of mass m in a gravitational field—as seen from a general Lorentz frame. They are *force equations*, with contortion (or torsion) playing the role of gravitational force.

By contraction with tetrads, and using identity (4.66), the equations of motion (6.50) and (6.51) can be written in a purely spacetime form, given respectively by

$$\frac{du_\mu}{ds} - \overset{\bullet}{\Gamma}{}^\rho{}_{\mu\nu} u_\rho u^\nu = -\overset{\bullet}{K}{}^\rho{}_{\mu\nu} u_\rho u^\nu \tag{6.52}$$

and

$$\frac{du^\mu}{ds} + \overset{\bullet}{\Gamma}{}^\mu{}_{\rho\nu} u^\rho u^\nu = \overset{\bullet}{K}{}^\mu{}_{\rho\nu} u^\rho u^\nu. \tag{6.53}$$

Comment 6.4 Using the contortion definition (4.69), the equation of motion (6.52) can be rewritten as

$$\frac{du_\mu}{ds} - \overset{\bullet}{\Gamma}{}^\rho{}_{\mu\nu} u_\rho u^\nu = \overset{\bullet}{T}{}^\rho{}_{\mu\nu} u_\rho u^\nu. \tag{6.54}$$

Substituting

$$\overset{\bullet}{T}{}^\rho{}_{\mu\nu} = \overset{\bullet}{\Gamma}{}^\rho{}_{\nu\mu} - \overset{\bullet}{\Gamma}{}^\rho{}_{\mu\nu}, \tag{6.55}$$

it reduces to

$$\frac{du_\mu}{ds} - \overset{\bullet}{\Gamma}{}^\rho{}_{\nu\mu} u_\rho u^\nu = 0. \tag{6.56}$$

Due to the "wrong" positions of the indices, however, as well as to the fact that the Weitzenböck connection is not symmetric in the last two indices, the left-hand side of the equation of motion (6.56) is not the covariant derivative of the four-velocity u_μ. That is to say, it is not the particle four-acceleration. This means that test particles do not follow the geodesics (or auto-parallels) of the "torsioned spacetime".

6.2.3 Equivalence with the Geodesic Equation

Let us rewrite the force equation (6.51) in the form

$$\frac{du^a}{ds} + (\overset{\bullet}{A}{}^a{}_{b\rho} - \overset{\bullet}{K}{}^a{}_{b\rho}) u^b u^\rho = 0. \tag{6.57}$$

Remembering that

$$\overset{\bullet}{A}{}^a{}_{b\rho} - \overset{\bullet}{K}{}^a{}_{b\rho} = \overset{\circ}{A}{}^a{}_{b\rho}, \tag{6.58}$$

with $\overset{\circ}{A}{}^a{}_{b\rho}$ the spin connection of General Relativity, the teleparallel force equation (6.51) is found to coincide with the geodesic equation

$$\frac{du^a}{ds} + \overset{\circ}{A}{}^a{}_{b\rho} u^b u^\rho = 0 \tag{6.59}$$

of General Relativity. We see in this way that the teleparallel description of the gravitational interaction is completely equivalent to the description of General Relativity.

There are conceptual differences, though. In General Relativity, a theory fundamentally based on the weak equivalence principle, curvature is used to *geometrize* the gravitational interaction. The gravitational interaction in this case is described by letting (spinless) particles to follow the curvature of spacetime. Geometry replaces the concept of force, and the trajectories are determined, not by force equations, but by geodesics. Teleparallel Gravity, on the other hand, attributes gravitation to torsion. Torsion, however, accounts for gravitation not by geometrizing the interaction, but by acting as a force. In consequence, there are no geodesics in Teleparallel Gravity, only force equations quite analogous to the Lorentz force equation of electrodynamics [1]. This is an expected result because, like electrodynamics, Teleparallel Gravity is a gauge theory.

Comment 6.5 The above equivalence should not be surprising because, when the invariant spacetime interval ds is written in the holonomic basis dx^μ, that is,

$$ds = \left(g_{\mu\nu}\, dx^\mu\, dx^\nu\right)^{1/2}, \tag{6.60}$$

the action functional

$$\mathcal{S} = -mc \int_{\mathsf{p}}^{\mathsf{q}} ds, \tag{6.61}$$

which is the same as (6.31), coincides with the usual general-relativistic form of the action, from where the geodesic equations are usually obtained.

6.3 Newtonian Limit

Through contraction with tetrads, and using Eq. (1.42), the force equation (6.48) can be rewritten in the purely spacetime form

$$\frac{du^\rho}{ds} + \dot{\Gamma}^\rho{}_{\mu\nu} u^\mu u^\nu = \dot{T}_\mu{}^\rho{}_\nu u^\mu u^\nu. \tag{6.62}$$

The newtonian limit is obtained by assuming that the gravitational field is stationary and weak. This means, respectively, that the time derivative of $B^a{}_\mu$ vanishes, and that $|B^a{}_\mu| \ll 1$. Accordingly, all particles are supposed to move with a sufficient small velocity so that the space components u^i of the four-velocity can be neglected in relation to the time component u^0. Under such conditions, the force equation (6.62) reduces to

$$\frac{du^\rho}{ds} + \dot{\Gamma}^\rho{}_{00} u^0 u^0 = \dot{T}_0{}^\rho{}_0 u^0 u^0. \tag{6.63}$$

In the class of frames in which the teleparallel spin connection $\dot{A}^a{}_{b\mu}$ vanishes, and choosing coordinates such that $\partial_\mu x^a = \delta^a_\mu$, the tetrad assumes the form

$$h^a{}_\mu = \delta^a_\mu + B^a{}_\mu. \tag{6.64}$$

This tetrad can be interpreted as an expansion up to first order in $B^a{}_\mu$. Up to this order, the Weitzenböck connection is [see Eq. (4.65)]

$$\dot{\Gamma}^\rho{}_{\mu\nu} = \partial_\nu B^\rho{}_\mu, \tag{6.65}$$

where we have used the identification $B^\rho{}_\mu = \delta^\rho_a B^a{}_\mu$. One can then verify that

$$\dot{\Gamma}^\rho{}_{00} = 0 \quad \text{and} \quad \dot{T}_0{}^\rho{}_0 = \partial^\rho B_{00}, \tag{6.66}$$

which brings Eq. (6.63) to the form

$$\frac{d^2x^\rho}{ds^2} = \partial^\rho B_{00} u^0 u^0. \tag{6.67}$$

Substituting

$$u^0 = c\frac{dt}{ds}, \tag{6.68}$$

it reduces to

$$\frac{d^2x^\rho}{ds^2} = c^2 \partial^\rho B_{00} \frac{dt^2}{ds^2}. \tag{6.69}$$

Since $\partial^0 B_{00} = 0$, the time component of this equation reads

$$\frac{d^2x^0}{ds^2} \equiv c\frac{d^2t}{ds^2} = 0, \tag{6.70}$$

whose solution says that dt/ds is a constant. The space components of Eq. (6.69) can then be written as

$$\frac{dv^j}{dt} = c^2 \partial^j B_{00}, \tag{6.71}$$

with $v^j = dx^j/dt$ the particle classical velocity. If we identify

$$c^2 B_{00} = \Phi, \tag{6.72}$$

with

$$\Phi = -\frac{GM}{|\boldsymbol{x}|} \tag{6.73}$$

the newtonian gravitational potential, we get

$$\frac{d\boldsymbol{v}}{dt} = -\boldsymbol{\nabla}\Phi, \tag{6.74}$$

where we have used that $\partial^i = -\partial_i$. This is the newtonian equation of motion. As expected due to the equivalence with the geodesic equation, the force equation of Teleparallel Gravity also has the correct newtonian limit.

Comment 6.6 It is interesting to observe that, since both teleparallel and newtonian gravity describe the gravitational interaction through a force, the newtonian limit follows much more naturally from Teleparallel Gravity than from General Relativity, where the concept of gravitational force is absent.

6.4 Gravitomagnetic Field

Let us consider the same limit of the previous section, but now keeping also terms linear in the space components u^i of the four-velocity. In this case, the force equation (6.62) assumes the form

$$\frac{du^\rho}{ds} + \dot{\Gamma}^\rho{}_{00} u^0 u^0 + (\dot{\Gamma}^\rho{}_{0i} + \dot{\Gamma}^\rho{}_{i0}) u^0 u^i = \dot{T}_0{}^\rho{}_0 u^0 u^0 + (\dot{T}_0{}^\rho{}_i + \dot{T}_i{}^\rho{}_0) u^0 u^i. \tag{6.75}$$

Substituting the connection (6.65), and discarding terms containing time-derivatives of the potential, this reduces to

$$\frac{d^2x^\rho}{ds^2} = \partial^\rho B_{00}u^0u^0 + \left[\partial^\rho(B_{0i} + B_{i0}) - \partial_i\left(B_0{}^\rho + B^\rho{}_0\right)\right]u^0u^i. \tag{6.76}$$

We see from this expression that, in the linear approximation, only the symmetric part of $B^\rho{}_\mu$ contributes to the equation of motion [2]. In this case Eq. (6.76) becomes

$$\frac{d^2x^\rho}{ds^2} = \partial^\rho B_{00}u^0u^0 + 2\left(\partial^\rho B_{0i} - \partial_i B_0{}^\rho\right)u^0u^i. \tag{6.77}$$

Substituting

$$u^0 = c\frac{dt}{ds} \quad \text{and} \quad u^i \equiv \frac{dx^i}{ds} = \frac{dx^i}{dt}\frac{dt}{ds} = v^i\frac{dt}{ds}, \tag{6.78}$$

and using the identification (6.72), it reduces to

$$\frac{d^2x^\rho}{ds^2} = \partial^\rho \Phi\frac{dt^2}{ds^2} + 2c\left(\partial^\rho B_{0i} - \partial_i B_0{}^\rho\right)\frac{dt^2}{ds^2}v^i. \tag{6.79}$$

Calling to mind that we are in the static-field limit and neglecting terms quadratic in the velocity v^i and its derivatives, the time component of this equation reads

$$\frac{d^2x^0}{ds^2} \equiv c\frac{d^2t}{ds^2} = 0, \tag{6.80}$$

where we have used that the velocity is orthogonal to the ordinary newtonian force: $v^i\partial_i\Phi = 0$. The solution to this equation implies that dt/ds equals a constant. As a consequence, the space components of Eq. (6.79) can be written in the form

$$\frac{d^2x^j}{dt^2} = -\partial_j\Phi + 2\left(\partial^j B_{0i} - \partial_i B_0{}^j\right)cv^i. \tag{6.81}$$

If we identify

$$\partial_j B_{0i} - \partial_i B_{0j} = \varepsilon_{jik}\frac{H^k}{c^2}, \tag{6.82}$$

we get [3]

$$\frac{d\boldsymbol{v}}{dt} = -\nabla\Phi - 2\frac{\boldsymbol{v}}{c}\times\boldsymbol{H}, \tag{6.83}$$

where we have used the identity $\partial^j = -\partial_j$. We see from this equation that H^k are the components of the gravitomagnetic field [4]. Since

$$\partial_j B_{0i} - \partial_i B_{0j} = \overset{\bullet}{T}{}_{0ji}, \tag{6.84}$$

Eq. (6.82) has the inverse

$$\frac{H^k}{c^2} = \tfrac{1}{2}\varepsilon^{kji}\overset{\bullet}{T}{}_{0ji}. \tag{6.85}$$

In Teleparallel Gravity, therefore, the gravitomagnetic field is written in terms of the potential B_{0i} in a way quite similar to the electromagnetic case, with (the zero-component of) torsion $\overset{\bullet}{T}{}_{0ji}$ replacing the electromagnetic field strength F_{ji}.

Comment 6.7 The action functional describing a *spinless* particle of momentum p_a in the presence of gravitation is

$$\mathscr{S} = \int_{\mathsf{p}}^{\mathsf{q}} \left[-h^a{}_\mu p_a\right] dx^\mu, \tag{6.86}$$

with

$$h^a{}_\mu = \partial_\mu x^a + \dot{A}^a{}_{b\mu} x^b + B^a{}_\mu \tag{6.87}$$

the tetrad field. If, in addition to momentum p_a, the particle has spin angular momentum density $s_{ab} = -s_{ba}$, its action functional, according to the teleparallel coupling prescription (5.33), turns out to be given by

$$\mathscr{S} = \int_{\mathsf{p}}^{\mathsf{q}} \left[-h^a{}_\mu p_a + \tfrac{1}{2}\left(\dot{A}^{ab}{}_\mu - \dot{K}^{ab}{}_\mu\right) s_{ab}\right] dx^\mu. \tag{6.88}$$

In Appendix A we show that the variation of this action functional gives rise to the teleparallel equivalent of the Papapetrou equation—and consequently to the Papapetrou equation itself.

6.5 Separating Inertial Effects from Gravitation

Take the tetrad field

$$h^a{}_\mu = \dot{\mathscr{D}}_\mu x^a + B^a{}_\mu. \tag{6.89}$$

Whereas the first term on the right-hand side is purely inertial, the second is purely gravitational. This means that both inertia and gravitation are included in $h^a{}_\mu$. As a consequence, the coefficient of anholonomy of h_a, which is defined by

$$f^c{}_{ab} = h_a{}^\mu h_b{}^\nu \left(\partial_\nu h^c{}_\mu - \partial_\mu h^c{}_\nu\right), \tag{6.90}$$

will also represent both inertia and gravitation. Of course, the same is true for the spin connection of General Relativity,

$$\mathring{A}^a{}_{bc} = \tfrac{1}{2}\left(f_b{}^a{}_c + f_c{}^a{}_b - f^a{}_{bc}\right). \tag{6.91}$$

In a local frame in which inertial effects exactly compensate gravitation, that connection vanishes,

$$\mathring{A}^a{}_{bc} \doteq 0, \tag{6.92}$$

and gravitation becomes locally undetectable. We recall that, according to the strong equivalence principle, this is possible at a point, or at most along a particle worldline. In this local frame, the geodesic equation (6.59) reduces to the equation of motion of a free particle:

$$\frac{du^a}{ds} = 0. \tag{6.93}$$

Let us consider again the identity

$$\mathring{A}^a{}_{b\rho} = \dot{A}^a{}_{b\rho} - \dot{K}^a{}_{b\rho}. \tag{6.94}$$

Since the teleparallel spin connection $\overset{\bullet}{A}{}^a{}_{b\rho}$ represents inertial effects only, this expression corresponds actually to a separation of gravitation from the inertial effects [5]. In the local frame in which the spin connection $\overset{\circ}{A}{}^a{}_{b\rho}$ vanishes, the identity (6.94) becomes

$$\overset{\bullet}{A}{}^a{}_{b\rho} \doteq \overset{\bullet}{K}{}^a{}_{b\rho}. \tag{6.95}$$

This expression shows explicitly that, in such a local frame, inertial effects (left-hand side) exactly compensates gravitation (right-hand side). Substituting (6.94) in the geodesic equation (6.59), we get back the force equation of Teleparallel Gravity:

$$\frac{du^a}{ds} + \overset{\bullet}{A}{}^a{}_{b\rho} u^b u^\rho = \overset{\bullet}{K}{}^a{}_{b\rho} u^b u^\rho. \tag{6.96}$$

The right-hand side represents the purely gravitational force, which transforms covariantly under local Lorentz transformations. The inertial forces coming from the frame non-inertiality are represented by the connection of the left-hand side, which is non-covariant by its very nature. In Teleparallel Gravity, therefore, whereas the gravitational effects are described by a covariant force, the non-inertial effects of the frame remain *geometrized* in the sense of General Relativity, and are represented by an inertia-related Lorentz connection. Notice that in the geodesic equation (6.59), both inertial and gravitational effects are described by the connection term of the left-hand side.

Comment 6.8 Although the inertial part of $\overset{\circ}{A}{}^a{}_{b\rho}$ does not contribute to some physical quantities, like curvature and torsion, it does contribute to others. One example is the energy-momentum density of gravitation, whose expression in General Relativity always include, in addition to the energy-momentum density of gravity itself, also the energy-momentum density of inertia, which is non-tensorial by definition. This is the reason why in General Relativity this density always shows up as a pseudotensor. In Chap. 10 we will examine this question in more detail.

In Teleparallel Gravity it is also possible to choose a *global* frame h'_b in which only the inertial effects vanish. In this frame $\overset{\bullet}{A}{}'^a{}_{b\rho} = 0$, and the equation of motion (6.96) assumes the purely gravitational form

$$\frac{du'^a}{ds} = \overset{\bullet}{K}{}'^a{}_{b\rho} u'^b u'^\rho. \tag{6.97}$$

In euclidean coordinates, the spacetime-indexed version of Eq. (6.97) is

$$\frac{du'^\mu}{ds} = \overset{\bullet}{K}{}'^\mu{}_{\nu\rho} u'^\nu u'^\rho. \tag{6.98}$$

In this form, the gravitational force becomes quite similar to the Lorentz force of electrodynamics. There is a difference, though: in contrast to the Lorentz force, which is linear in the four-velocity, the gravitational force is quadratic in the four-velocity. This difference is related to the strictly attractive character of gravitation.

Comment 6.9 Since no inertial effects are present in the above equations of motion, we can conclude that Teleparallel Gravity does not need to use the equivalence principle to describe the gravitational interaction [6]. In Chap. 11 this property will be explored in more detail.

6.6 A Genuine Gravitational Connection

Due to the fact that General Relativity is deeply rooted on the equivalence principle, its spin connection involves both gravitation and inertial effects. As a consequence, it is always possible to find a local frame in which inertial effects exactly compensate gravitation, in such a way that the connection vanishes at a point, or along a trajectory:

$$\overset{\circ}{A}{}^{a}{}_{b\rho} \doteq 0. \tag{6.99}$$

Since we know there is gravitation at that point, such connection cannot be a genuine gravitational variable in the usual sense of field theory. Notice, in particular, that any approach to quantum gravity based on this connection will necessarily include a quantization of the inertial forces—whatever that may come to mean. Considering furthermore the divergent asymptotic behavior of the inertial effects, such approach will likely face additional difficulties.

There is an additional problem: the connection behavior of $\overset{\circ}{A}{}^{a}{}_{b\rho}$ is due to its inertial content, not to gravitation itself. This can be seen from the decomposition

$$\overset{\circ}{A}{}^{a}{}_{b\rho} = \overset{\bullet}{A}{}^{a}{}_{b\rho} - \overset{\bullet}{K}{}^{a}{}_{b\rho}, \tag{6.100}$$

where the first term on the right-hand side represents its inertial, non-covariant part, whereas the second represents its gravitational part, which is a tensor. This means that it is not a gravitational connection—its gravitational content is covariant—but an inertial connection. One should not expect, therefore, any dynamical effect coming from a "gaugefication" of the Lorentz group. As a matter of fact, local Lorentz transformations are related with different classes of non-inertial frames. Due to the local equivalence between inertial effects and gravitation, they are related to the strong equivalence principle, and consequently to the geometric description of General Relativity, not to a gauge description. They are, in this sense, similar to diffeomorphism, another symmetry empty of dynamical meaning.

In Teleparallel Gravity, on the other hand, the gravitational field is not represented by a Lorentz connection, but by a translational-valued gauge potential $B^{a}{}_{\mu}$, the non-trivial part of the tetrad field, as can be seen from Eq. (6.89). In this theory, Lorentz connections keep their special relativistic role, representing inertial effects only. Considering that the translational gauge potential represents gravitation only, to the exclusion of inertial effects, it is a genuine gravitational connection, and consequently the natural field-variable to be quantized in any approach to quantum gravity.

References

1. de Andrade, V.C., Pereira, J.G.: Phys. Rev. D **56**, 4689 (1997). gr-qc/9703059
2. Obukhov, Yu.N., Pereira, J.G.: Phys. Rev. D **67**, 044008 (2003). gr-qc/0212078
3. Mashhoon, B.: Gravitoelectromagnetism: a brief review. In: Iorio, L. (ed.) The Measurement of Gravitomagnetism: A Challenging Enterprise. Nova Science, New York (2007). gr-qc/0311030

4. Ciufolini, I., Wheeler, J.A.: Gravitation and Inertia. Princeton University Press, Princeton (1995)
5. Aldrovandi, R., Guillen, L.C.T., Pereira, J.G., Vu, K.H.: Bringing together gravity and the quanta. Contribution to the Proceedings of the Albert Einstein Century International Conference, Paris, 18–22 July 2005. gr-qc/0603122
6. Aldrovandi, R., Pereira, J.G., Vu, K.H.: Gen. Relativ. Gravit. **36**, 101 (2004). gr-qc/0304106

Chapter 7
Global Formulation for Gravity

Due to their shared abelian gauge structure, Teleparallel Gravity and Electromagnetism are similar in several aspects. By analogy to the phase-factor approach to Maxwell's theory, a teleparallel non-integrable phase-factor formalism for gravitation can be developed. It represents the quantum mechanical version of the classical gravitational force, and leads to simple descriptions of the Colella-Overhauser-Werner experiment and of the gravitational Aharonov-Bohm effect. In the classical (non-quantum) limit, it reduces to the force equation of Teleparallel Gravity.

7.1 Phase Factor Approach

As is widely known, in addition to the usual *differential* formalism, electromagnetism (as gauge theories in general) has a *global* formulation in terms of a non-integrable phase factor [1, 2]. According to that approach, electromagnetism can be considered as the gauge invariant effect of a non-integrable (that is, path-dependent) phase factor. For a particle with electric charge q traveling from an initial point p to a final point q, the phase factor is given by

$$\Phi_e(\mathsf{p}|\mathsf{q}) = \exp\left(\frac{iq}{\hbar c}\int_{\mathsf{p}}^{\mathsf{q}} A_\mu \, dx^\mu\right), \tag{7.1}$$

where A_μ is the electromagnetic gauge potential. In the classical (non-quantum) limit, the non-integrable phase factor approach yields the same results as those obtained from the Lorentz force equation

$$\frac{du^\mu}{d\sigma} = \frac{q}{mc^2} F^\mu{}_\nu u^\nu. \tag{7.2}$$

In this sense, the phase-factor approach can be considered the *quantum* generalization of the *classical* Lorentz force equation. It is actually more general, as it can be used both on simply-connected and on multiply-connected domains. Its use is mandatory, for example, to describe the Aharonov-Bohm effect, a quantum phenomenon taking place in a multiply-connected space. While a differential equation

R. Aldrovandi, J.G. Pereira, *Teleparallel Gravity*, Fundamental Theories of Physics 173,
DOI 10.1007/978-94-007-5143-9_7,

as (7.2) is strictly local, an integrated object as the phase in (7.1) is global, and consequently able to take topological effects into account.

By its similarity to A_μ, the teleparallel gauge potential $B^a{}_\mu$ can be used to provide a global formulation for gravitation [3]. To start with, let us notice that the electromagnetic phase factor (7.1) has the form

$$\Phi_e(\mathsf{p}|\mathsf{q}) = \exp\left(-\frac{i}{\hbar}\mathscr{S}_e\right), \tag{7.3}$$

where

$$\mathscr{S}_e = -\frac{q}{c}\int_\mathsf{p}^\mathsf{q} A_\mu \, dx^\mu \tag{7.4}$$

is the *interaction part* of the action integral for a charged particle within an electromagnetic field [see Eq. (6.35)]. We can similarly write the gravitational phase factor as

$$\Phi_g(\mathsf{p}|\mathsf{q}) = \exp\left(-\frac{i}{\hbar}\mathscr{S}_g\right), \tag{7.5}$$

where $\mathscr{S}_g$ is the *interaction part* of the action integral for a particle of mass m in a gravitational field. From Eq. (6.33) we see that such a part is given by

$$\mathscr{S}_g = -mc\int_\mathsf{p}^\mathsf{q} \left(-\overset{\bullet}{A}{}^a{}_{b\mu} x^b + B^a{}_\mu u_a\right) dx^\mu . \tag{7.6}$$

The first term of the integrand represents the interaction of the particle with the inertial effects present in the frame, and the second the gravitational interaction. As we are interested only in the gravitational interaction, we can choose to work in an inertial frame, in which $\overset{\bullet}{A}{}^a{}_{b\mu} = 0$. In this case, the action integral reduces to

$$\mathscr{S}_g = -mc\int_\mathsf{p}^\mathsf{q} B^a{}_\mu u_a \, dx^\mu, \tag{7.7}$$

and the gravitational phase factor assumes the form

$$\Phi_g(\mathsf{p}|\mathsf{q}) = \exp\left(\frac{imc}{\hbar}\int_\mathsf{p}^\mathsf{q} B^a{}_\mu u_a \, dx^\mu\right). \tag{7.8}$$

Just as happens for the electromagnetic phase factor, it represents the *quantum* mechanical law that replaces the *classical* gravitational Lorentz force equation (6.97).

7.2 Colella-Overhauser-Werner Experiment

As a first application of the gravitational non-integrable phase factor (7.8), we consider the Colella-Overhauser-Werner (COW) experiment [4]. It consists of using a neutron interferometer to observe the quantum mechanical phase shift of neutrons caused by their interaction with Earth's gravitational field, assumed to be newtonian. As seen in Sect. 6.3, a newtonian gravitational field is characterized by the

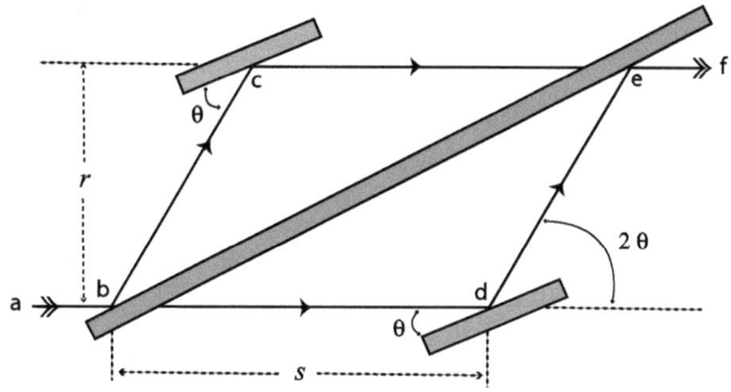

Fig. 7.1 Schematic illustration of the COW neutron interferometer

condition that only $B_{00} \neq 0$. Furthermore, as the experience is performed with thermal neutrons, it is possible to use the small velocity approximation, according to which

$$u^0 = \gamma \equiv \left[1 - \left(v^2/c^2\right)\right]^{-1/2} \simeq 1. \tag{7.9}$$

When acting on the wave function of such particles, therefore, the gravitational phase factor (7.8) assumes the form

$$\Phi_g(\mathsf{p}|\mathsf{q}) = \exp\left(\frac{im}{\hbar}\int_{\mathsf{p}}^{\mathsf{q}} c^2 B_{00}\, dt\right). \tag{7.10}$$

According to Eq. (6.72), in the newtonian approximation the term $c^2 B_{00}$ is identified with the (assumed homogeneous) Earth newtonian potential, which is here given by

$$c^2 B_{00} = gz. \tag{7.11}$$

In this expression, g is the gravitational acceleration, supposed not to change significantly in the domain of the experience, and z is the distance from Earth taken from some reference point. Consequently, the phase factor can be rewritten in the form

$$\Phi_g(\mathsf{p}|\mathsf{q}) = \exp\left(\frac{img}{\hbar}\int_{\mathsf{p}}^{\mathsf{q}} z(t)\, dt\right) \equiv \exp i\varphi. \tag{7.12}$$

Let us now compute the phase φ through the two trajectories of Fig. 7.1. First, we consider the trajectory bde. Assuming that the segment bd is at $z = 0$, we obtain

$$\varphi_{\mathsf{bde}} = \frac{mg}{\hbar}\int_{\mathsf{d}}^{\mathsf{e}} z(t)\, dt. \tag{7.13}$$

For the trajectory bce, we have

$$\varphi_{\mathsf{bce}} = \frac{mg}{\hbar}\int_{\mathsf{b}}^{\mathsf{c}} z(t)\, dt + \frac{mgr}{\hbar}\int_{\mathsf{c}}^{\mathsf{e}} dt. \tag{7.14}$$

As the phase contribution along the segments de and bc are equal, they cancel out from the phase difference

$$\Delta\varphi \equiv \varphi_{\mathsf{bce}} - \varphi_{\mathsf{bde}} = \frac{mgr}{\hbar}\int_{\mathsf{c}}^{\mathsf{e}} dt. \tag{7.15}$$

Since the neutron velocity is constant along the segment ce, the integral is

$$\int_{\mathsf{c}}^{\mathsf{e}} dt \equiv \frac{s}{v} = \frac{sm\lambda}{h}, \tag{7.16}$$

where s is the length of the segment ce, and $\lambda = h/mv$ is the de Broglie wavelength associated with the neutron. We thus obtain

$$\Delta\varphi = s\frac{2\pi g r \lambda m^2}{h^2}, \tag{7.17}$$

which is the gravitationally-induced phase difference of the COW experiment [4].

Comment 7.1 The dependence of the phase difference on the neutron mass seems to indicate that, at the quantum level, gravitation is no longer universal [5, 6]. If a quantum version of the equivalence principle is introduced [7, 8], however, it is possible to eliminate the mass dependence of the COW effect, keeping thus gravitation universal also at the quantum level—at least for the case of the COW experiment.

7.3 Gravitational Aharonov-Bohm Effect

As a second application we use the phase factor (7.8) to study the gravitational analog of the Aharonov-Bohm effect [9–11]. The usual (electromagnetic) Aharonov-Bohm effect consists in a shift, by a constant amount, of the electron interferometry wave pattern, in a region where there is no magnetic field, but there is a nontrivial electromagnetic gauge potential A_i [12]. Analogously, the gravitational Aharonov-Bohm effect will consist in a similar shift of the same wave pattern, but produced by the presence of a gravitational gauge potential B_{0i}.

Comment 7.2 As the phase shift of the COW experiment is produced by the coupling of the neutron mass with the component B_{00} of the translational gauge potential, it can be considered as a gravitoelectric Aharonov-Bohm effect. A similar denomination is used in the electromagnetic case [13].

Phenomenologically, this kind of effect might be present near a massive, rapidly rotating source, like a neutron star for example. Of course, differently from an ideal apparatus, in a real situation the gravitational field cannot be eliminated, and consequently the gravitational Aharonov-Bohm effect should be added to the other effects also causing a phase change.

Let us consider first the case in which there is no gravitational field at all. If the electrons are emitted with a characteristic momentum p, then its wave function has the de Broglie wavelength $\lambda = h/p$. Denoting by L the distance between the slits and the screen (see Fig. 7.2), and by d the distance between the two slits, when the conditions

$$L \gg \lambda, \qquad L \gg x, \qquad L \gg d$$

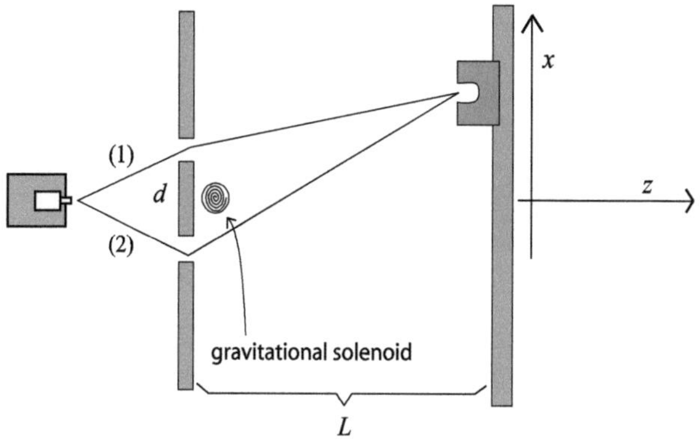

Fig. 7.2 Schematic view of the gravitational Aharonov-Bohm electron interferometer

are satisfied the phase difference at a distance x from the central point of the screen is given by

$$\delta^0\varphi(x) = \frac{2\pi x d}{L\lambda}. \tag{7.18}$$

This expression defines the wave pattern on the screen.

We consider now the case in which a kind of infinite *gravitational solenoid* produces a purely static gravitomagnetic field flux concentrated in its interior, as shown in Fig. 7.2. In the ideal situation, the gravitational field outside the solenoid vanishes completely, but there is a nontrivial gauge potential B_{0i}. When we let the electrons to move outside the solenoid, phase factors corresponding to paths lying on one side of the solenoid will interfere with phase factors corresponding to paths lying on the other side, which will produce an additional phase shift at the screen. Let us then calculate this additional phase shift. The gravitational phase factor (7.8) for the physical situation above described is

$$\Phi_g(\mathsf{p}|\mathsf{q}) = \exp\left(-\frac{imc}{\hbar}\int_{\mathsf{p}}^{\mathsf{q}} u^0 \boldsymbol{B}_0 \cdot d\boldsymbol{r}\right), \tag{7.19}$$

where $\boldsymbol{B}_0$ is the vector with components ${B_0}^i = -B_{0i}$. Since

$$u^0 = \gamma \equiv \left[1 - \left(v^2/c^2\right)\right]^{-1/2},$$

and considering a constant electron velocity v, we can write

$$\Phi_g(\mathsf{p}|\mathsf{q}) = \exp\left(-\frac{i\gamma mc}{\hbar}\int_{\mathsf{p}}^{\mathsf{q}} \boldsymbol{B}_0 \cdot d\boldsymbol{r}\right). \tag{7.20}$$

Now, denoting by φ_1 the phase corresponding to a path lying on one side of the solenoid, and by φ_2 the phase corresponding to a path lying on the other side, the phase difference at the screen will be

$$\delta\varphi \equiv \varphi_2 - \varphi_1 = \frac{\gamma mc}{\hbar}\oint \boldsymbol{B}_0 \cdot d\boldsymbol{r}. \tag{7.21}$$

Using Stokes theorem, we get

$$\delta\varphi = \frac{\gamma mc}{\hbar} \oint (\nabla \times \boldsymbol{B}_0) \cdot d\boldsymbol{\sigma}, \tag{7.22}$$

or equivalently

$$\delta\varphi = \frac{\mathscr{E}\Omega}{\hbar c}, \tag{7.23}$$

where $\mathscr{E} = \gamma mc^2$ is the electron kinetic energy, and

$$\Omega = \oint (\nabla \times \boldsymbol{B}_0) \cdot d\boldsymbol{\sigma} \equiv \frac{1}{c^2} \oint \boldsymbol{H} \cdot d\boldsymbol{\sigma} \tag{7.24}$$

is the flux of gravitomagnetic field $\boldsymbol{H}$ inside the solenoid. In components, the gravitomagnetic field $\boldsymbol{H}$ is written as

$$\frac{H^i}{c^2} = \tfrac{1}{2}\varepsilon^{ijk}(\partial_j B_{0k} - \partial_k B_{0j}) = \tfrac{1}{2}\varepsilon^{ijk} \overset{\bullet}{T}{}_{0jk}. \tag{7.25}$$

Expression (7.23) gives the phase difference produced by the interaction of the particle kinetic energy with a gauge potential, which gives rise to the gravitational Aharonov-Bohm effect. As this phase difference is written in terms of the energy, it applies equally to massive and massless particles. There is a difference, though: whereas for massive particles it is a genuine quantum effect, for massless particles, due to their intrinsic wave character, it can be considered as a classical effect. In fact, for $\mathscr{E} = \hbar\omega$, the phase difference becomes

$$\delta\varphi = \frac{\omega\Omega}{c}, \tag{7.26}$$

which does not depend on the Planck constant.

It is important to remark that, like in the case of the electromagnetic Aharonov-Bohm effect, whose phase shift is given by

$$\delta_e\varphi = \frac{q\Omega_e}{\hbar}, \tag{7.27}$$

with Ω_e the magnetic flux inside the solenoid, the gravitational phase shift (7.23) is independent of the position x on the screen, and consequently the wave pattern will be wholly shifted by a constant amount. In contrast to the electromagnetic case, however, where the phase difference depends only on fundamental properties of the particle, in the gravitational case it depends on the particle kinetic energy, which in turn depends on the particle velocity and mass. Due to this property, it can be shown that, whereas in the electromagnetic case it is possible to define a *quantum* of magnetic flux, in the gravitational case, owing to the dependence of the phase shift on the electron kinetic energy, it is not possible to define a particle-independent *quantum* of gravitomagnetic flux [14].

7.4 Quantum Versus Classical Approaches

We proceed now to show that, in the classical limit, the non-integrable phase factor approach reduces to the usual approach provided by the gravitational force equation.

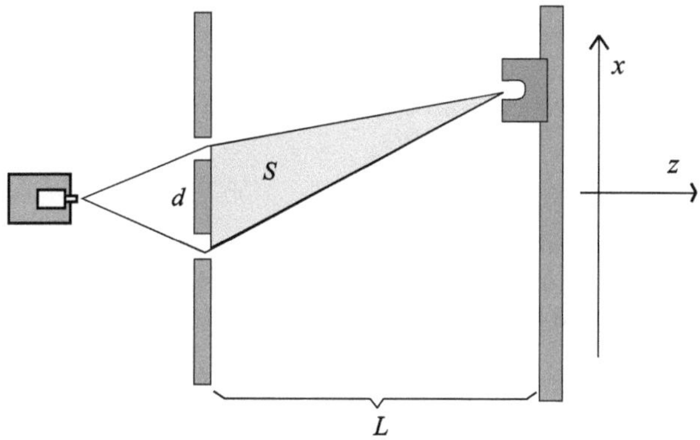

Fig. 7.3 Schematic illustration of the electron interference experiment in the presence of a gravitomagnetic field permeating the whole region between the slits and the screen. The only contribution to the phase shift comes from the flux inside the surface S delimited by the two trajectories

In electromagnetism, the standard argument is well-known: the phase shift turning up in the quantum case is, in the classical limit, the Lorentz force equation. We intend, however, to illustrate the result directly, and for that we consider again the electron interferometry slit experiment, this time with a homogeneous static gravitomagnetic field $\boldsymbol{H}$ permeating the whole region between the slits and the screen (see Fig. 7.3). This phase shift, according to Eq. (7.23), is given by $\mathscr{E}\Omega/\hbar c$, with Ω the flux through the surface S circumscribed by the two trajectories. This field is supposed to point in the negative y-direction, and will produce a phase shift which is to be added to the ordinary phase shift

$$\delta^0\varphi(x) = \frac{2\pi x d}{L\lambda}, \tag{7.28}$$

extant in the absence of gravitomagnetic field. It is easily seen that

$$S = \frac{Ld}{2} \tag{7.29}$$

for any value of x. The flux is, consequently,

$$\Omega = -\frac{H_y L d}{2c^2}, \tag{7.30}$$

where $\boldsymbol{H} = -H_y \hat{e}_y$, with $\hat{e}_y$ a unity vector in the y direction. Therefore, the total phase difference will be

$$\delta\varphi = \frac{2\pi x d}{L\lambda} - \frac{\mathscr{E} H_y L d}{2\hbar c^3}. \tag{7.31}$$

This is the total phase-shift yielded by the phase-factor approach.

In the classical limit, the slit experiment can be interpreted in the following way. The electrons traveling through the gravitomagnetic field have their movement direction changed. This means that they are subjected to a force in the x-direction [15]. For $x \ll L$, we can approximately write the electrons velocity as $v \simeq v_z$. In this case, they will be transversally accelerated by the gravitomagnetic field during the time interval

$$\Delta t = \frac{L}{v_z}. \tag{7.32}$$

This transversal x-acceleration is given by

$$a_x = \frac{2x}{(\Delta t)^2}. \tag{7.33}$$

Since the attained acceleration is constant, we can choose a specific point to calculate it. Let us then consider the point of maximum intensity on the screen, which is determined by the condition $\delta\varphi = 0$. This yields

$$x = \frac{H_y \lambda L^2 \mathscr{E}}{2hc^3}. \tag{7.34}$$

The acceleration is then found to be

$$a_x = \frac{H_y \lambda \mathscr{E} v_z^2}{hc^3}. \tag{7.35}$$

From the classical point of view, therefore, we can say that the electrons experience a force in the x-direction given by

$$\mathscr{F}_x \equiv \gamma m a_x = \frac{\mathscr{E} v_z H_y}{c^3} \frac{\lambda p}{h}, \tag{7.36}$$

with $p = \gamma m v_z$ the electron momentum. Using the de Broglie relation $\lambda = h/p$, the Planck constant is eliminated, and we get the classical result

$$\mathscr{F}_x = \frac{\mathscr{E}}{c^3} v_z H_y = \frac{\mathscr{E}}{c^3} (\boldsymbol{v} \times \boldsymbol{H})_x. \tag{7.37}$$

Comment 7.3 This force is quite similar to the electromagnetic Lorentz force, with the kinetic energy $\mathscr{E} = \gamma mc^2$ replacing the electric charge, and $\boldsymbol{H}/c^2$ replacing the usual magnetic field.

The equation of motion corresponding to the force (7.37) is

$$\frac{\overset{\bullet}{\nabla} p_x}{\nabla t} = \frac{\mathscr{E}}{c^3} (\boldsymbol{v} \times \boldsymbol{H})_x, \tag{7.38}$$

where $\overset{\bullet}{\nabla}/\nabla t$ represents a time covariant derivative in the Weitzenböck connection. Using $p^i = \gamma m v^i$, it can be rewritten as

$$\frac{\overset{\bullet}{\nabla} v_i}{\nabla t} = \frac{1}{c} (\boldsymbol{v} \times \boldsymbol{H})_i \equiv \frac{1}{c} \varepsilon_{ijk} v^j H^k. \tag{7.39}$$

From Eq. (7.25), however, we see that $\varepsilon_{ijk} H^k = c^2 \overset{\bullet}{T}_{0ij}$. In terms of torsion components, therefore, we have

$$\frac{\overset{\bullet}{\nabla} v_i}{\nabla t} = c \overset{\bullet}{T}_{0ij} v^j. \tag{7.40}$$

Furthermore, a non-vanishing gravitomagnetic field $B^0{}_i$ does not change the time components of the spacetime metric g_{00}, which remains that of a Minkowski spacetime. This means that $dt = (\gamma/c)\, ds$, and consequently we can write

$$\frac{\overset{\bullet}{\nabla} v_i}{\nabla s} = \gamma \overset{\bullet}{T}{}_{0ij} v^j. \tag{7.41}$$

Using the relations $u^0 = \gamma$ and $u^j = (\gamma v^j)/c$, as well as the fact that the gravitomagnetic field does not change the absolute value of the particle velocity, and consequently $d\gamma/ds = 0$, we obtain

$$\frac{\overset{\bullet}{\nabla} u_i}{\nabla s} = \overset{\bullet}{T}{}_{0ij} u^0 u^j. \tag{7.42}$$

This force equation is a particular case of the gravitational Lorentz force equation [see Eq. (6.54)],

$$\frac{\overset{\bullet}{\nabla} u_\mu}{\nabla s} \equiv \frac{du_\mu}{ds} - \overset{\bullet}{\Gamma}{}^\lambda{}_{\mu\rho} u_\lambda u^\rho = \overset{\bullet}{T}{}_{\lambda\mu\rho} u^\lambda u^\rho, \tag{7.43}$$

for the case in which only the gravitomagnetic component $\boldsymbol{H}$ is non-vanishing. We see in this way that the phase factor approach reduces, in the classical limit, to the gravitational Lorentz force equation, which is equivalent to the geodesic equation of General Relativity [as discussed in Sect. 6.2.3].

References

1. Yang, C.N.: Phys. Rev. Lett. **33**, 445 (1974)
2. Wu, T.T., Yang, C.N.: Phys. Rev. D **12**, 3845 (1975)
3. Aldrovandi, R., Pereira, J.G., Vu, K.H.: Class. Quantum Gravity **21**, 51 (2004). gr-qc/0310110
4. Colella, R., Overhauser, A.W., Werner, S.A.: Phys. Rev. Lett. **34**, 1472 (1974)
5. Greenberger, D.: Ann. Phys. **47**, 116 (1968)
6. Greenberger, D., Overhauser, A.W.: Rev. Mod. Phys. **51**, 43 (1979)
7. Lämmerzahl, C.: Gen. Relativ. Gravit. **28**, 1043 (1996). gr-qc/9605065
8. Lämmerzahl, C.: Acta Phys. Pol. **29**, 1057 (1998). gr-qc/9807072
9. Lawrence, A.K., Leiter, D., Samozi, G.: Nuovo Cimento B **17**, 113 (1973)
10. Ford, L.H., Vilenkin, A.: J. Phys. A **14**, 2353 (1981)
11. Bezerra, V.B.: Class. Quantum Gravity **8**, 1939 (1991)
12. Aharonov, Y., Bohm, D.: Phys. Rev. **115**, 485 (1959)
13. Peshkin, M., Tonomura, A.: The Aharonov-Bohm Effect. Lecture Notes in Physics, vol. 340. Springer, Berlin (1989)
14. Harris, E.G.: Am. J. Phys. **64**, 378 (1996)
15. Felsager, B.: Geometry, Particles, and Fields. Springer, New York (1998)

Chapter 8
Hodge Dual for Soldered Bundles

To account for all possible contractions allowed by the presence of the solder form, a generalized Hodge dual must be defined in the case of soldered bundles. Although for curvature the generalized dual coincides with the usual one, for torsion it gives a new dual definition.

8.1 Why a New Dual

Let Ω^p be the space of p-forms on an n-dimensional manifold $\mathbb{R}$ with metric $g_{\mu\nu}$. Since vector spaces Ω^p and Ω^{n-p} have the same finite dimension, they are isomorphic. The presence of a metric renders it possible to single out an unique isomorphism between them, called Hodge dual. Using a coordinate basis, the Hodge dual of a p-form $\alpha \in \Omega^p$,

$$\alpha = \frac{1}{p!}\alpha_{\mu_1\ldots\mu_p} dx^{\mu_1} \wedge \cdots \wedge dx^{\mu_p}, \tag{8.1}$$

is the $(n-p)$-form $\star\alpha \in \Omega^{n-p}$ defined by [1]

$$\star\alpha = \frac{h}{(n-p)!p!}\varepsilon_{\mu_1\mu_2\ldots\mu_n}\alpha^{\mu_1\ldots\mu_p} dx^{\mu_{p+1}} \wedge \cdots \wedge dx^{\mu_n}, \tag{8.2}$$

where, we recall,

$$h = \det\left(h^a{}_\mu\right) = \sqrt{-g}, \tag{8.3}$$

with $g = \det(g_{\mu\nu})$, and $\varepsilon_{\mu_1\mu_2\ldots\mu_n}$ is the totally anti-symmetric Levi-Civita symbol discussed in Sect. 1.7. The operator $\star$ satisfies the property

$$\star\star\alpha = (-1)^{p(n-p)+(n-s)/2}\alpha, \tag{8.4}$$

where s is the metric signature. Its inverse is, consequently,

$$\star^{-1} = (-1)^{p(n-p)+(n-s)/2}\star. \tag{8.5}$$

This dual operator can be used to define an inner product on Ω^p, given by

$$\alpha \wedge \star\beta = \langle\alpha, \beta\rangle vol, \tag{8.6}$$

R. Aldrovandi, J.G. Pereira, *Teleparallel Gravity*, Fundamental Theories of Physics 173, DOI 10.1007/978-94-007-5143-9_8, © Springer Science+Business Media Dordrecht 2013

where *vol* is the volume element. Conversely, given an inner product, Eq. (8.6) can be used to define the dual operator.

For non-soldered bundles, the dual operator can be generalized to act on vector-valued p-forms in a straightforward way. Let β be a vector-valued p-form on the n-dimensional base space $\mathbb{R}$, taking values on a vector space $\mathbb{V}$,

$$\beta = J_A \beta^A, \tag{8.7}$$

where the set $\{J_A\}$ is a basis for the vector space $\mathbb{V}$. Its dual is the vector-valued $(n-p)$-form

$$\star\beta = \frac{h}{(n-p)!p!} \varepsilon_{\mu_1\mu_2\ldots\mu_n} J_A \beta^{A\mu_1\ldots\mu_p} dx^{\mu_{p+1}} \wedge \cdots \wedge dx^{\mu_n}. \tag{8.8}$$

The components $\beta^{A\mu_1\ldots\mu_p}$ bear now an internal space index, which is not related to the external indices μ_i. An example of such form is the Yang-Mills field strength in a four-dimensional spacetime

$$F = \tfrac{1}{2} J_A F^A{}_{\mu\nu} dx^\mu \wedge dx^\nu, \tag{8.9}$$

a 2-form assuming values in the Lie algebra of the unitary group $SU(n)$ [see Chap. 3]. Its dual will have components

$$\star F^A{}_{\mu\nu} = \frac{h}{2} \varepsilon_{\mu\nu\rho\sigma} F^{A\rho\sigma}. \tag{8.10}$$

For soldered bundles [2], on the other hand, *due to the presence of the solder form, internal and external indices can be transformed into each other, and this feature leads to the possibility of defining a new dual operator on Ω^p* [3]. The main requirement of this new definition is that, since (8.4) is valid for p-forms on soldered bundles, we want to make it true also for vector-valued p-forms. As usual, the inner product $\langle\alpha, \beta\rangle$ of two vector-valued p-forms α and β will be defined by taking the trace in the algebraic indices:

$$\mathrm{Tr}(\alpha \wedge \star\beta) = \langle\alpha, \beta\rangle vol. \tag{8.11}$$

For definiteness, we are going to obtain the new dual Hodge definition for the specific cases of torsion and curvature, quantities related to connections living in soldered bundles.

8.2 Dual Torsion

As we have seen in Sect. 1.3, torsion is a 2-form assuming values in the Lie algebra of the translation group,

$$\boldsymbol{T} = \tfrac{1}{2} T^a{}_{\mu\nu} P_a \, dx^\mu \wedge dx^\nu, \tag{8.12}$$

with $P_a = \partial_a$ the translation generators. Differently from the field strength of internal (non-soldered) gauge theories, in soldered bundles algebraic and spacetime indices can be transformed into each other through the use of the tetrad field. In fact, through contraction with the tetrad, it can be rewritten in a purely spacetime form:

$$T^{\lambda}{}_{\mu\nu} = h_a{}^{\lambda} T^a{}_{\mu\nu}. \tag{8.13}$$

This property opens up the possibility of new contractions in the dual definition (8.2). For the case of torsion, a general Hodge dual involving all possible index contractions is of the form

$$\star T^{\lambda}{}_{\mu\nu} = h\varepsilon_{\mu\nu\rho\sigma}\big(aT^{\lambda\rho\sigma} + bT^{\rho\lambda\sigma} + cT^{\theta\rho}{}_{\theta}g^{\lambda\sigma}\big), \tag{8.14}$$

with a, b, c constant coefficients. As a matter of fact, two coefficients suffice to define the generalized dual torsion. To see that, let us take the first and the second terms of (8.14):

$$a\varepsilon_{\mu\nu\rho\sigma}T^{\lambda\rho\sigma} \quad \text{and} \quad b\varepsilon_{\mu\nu\rho\sigma}T^{\rho\lambda\sigma}. \tag{8.15}$$

Since $T^{\lambda\rho\sigma}$ is anti-symmetric in $\rho\sigma$, whereas $T^{\rho\lambda\sigma}$ has no definite symmetry in $\rho\sigma$, and considering that both are contracted with $\varepsilon_{\mu\nu\rho\sigma}$, the first term contributes with half the number of independent terms in relation to the second term. This means that, in order to eliminate equivalent contractions, it is necessary that the coefficient a be half the value of b. We then set $b = 2a$, which yields

$$\star T^{\lambda}{}_{\mu\nu} = h\varepsilon_{\mu\nu\rho\sigma}\big(aT^{\lambda\rho\sigma} + 2aT^{\rho\lambda\sigma} + cT^{\theta\rho}{}_{\theta}g^{\lambda\sigma}\big). \tag{8.16}$$

Now, in a four-dimensional spacetime with metric signature $s = 2$ the dual torsion must, as any 2-form, satisfy the relation

$$\star\star T^{\rho}{}_{\mu\nu} = -T^{\rho}{}_{\mu\nu}. \tag{8.17}$$

This condition yields the following algebraic system:

$$8a^2 - 2ac = 1 \tag{8.18}$$

$$8a^2 + 2ac = 0. \tag{8.19}$$

There are two solutions, which differ by a global sign:

$$a = 1/4 \qquad c = -1 \tag{8.20}$$

and

$$a = -1/4 \qquad c = 1. \tag{8.21}$$

Choosing the solution with $a > 0$, the generalized dual torsion is found to be

$$\star T^{\rho}{}_{\mu\nu} = h\varepsilon_{\mu\nu\alpha\beta}\big(\tfrac{1}{4}T^{\rho\alpha\beta} + \tfrac{1}{2}T^{\alpha\rho\beta} - T^{\lambda\alpha}{}_{\lambda}g^{\rho\beta}\big). \tag{8.22}$$

Defining the tensor

$$S^{\rho\alpha\beta} = -S^{\rho\beta\alpha} := K^{\alpha\beta\rho} - g^{\rho\beta}T^{\sigma\alpha}{}_{\sigma} + g^{\rho\alpha}T^{\sigma\beta}{}_{\sigma}, \tag{8.23}$$

with

$$K^{\alpha\beta\rho} = \tfrac{1}{2}\big(T^{\beta\alpha\rho} + T^{\rho\alpha\beta} - T^{\alpha\beta\rho}\big) \tag{8.24}$$

the contortion tensor [see Eq. (1.67)], the dual torsion can be written in the compact form [3]

$$\star T^{\rho}{}_{\mu\nu} = \frac{h}{2} \varepsilon_{\mu\nu\alpha\beta} S^{\rho\alpha\beta}. \tag{8.25}$$

Multiplying both sides by $h^a{}_\rho$, it becomes

$$\star T^{a}{}_{\mu\nu} = \frac{h}{2} \varepsilon_{\mu\nu\alpha\beta} S^{a\alpha\beta}. \tag{8.26}$$

This is the generalized expression for the Hodge dual torsion.

8.3 Dual Curvature

As we have seen in Sect. 1.3, curvature is a 2-form assuming values in the Lie algebra of the Lorentz group,

$$\boldsymbol{R} = \tfrac{1}{4} R^a{}_{b\mu\nu} S_a{}^b \, dx^\mu \wedge dx^\nu, \tag{8.27}$$

with $S_a{}^b$ the Lorentz generators. Through contraction with the tetrad, it can be rewritten in a purely spacetime form:

$$R^\alpha{}_{\beta\mu\nu} = h_a{}^\alpha h^b{}_\beta R^a{}_{b\mu\nu}. \tag{8.28}$$

Hence, analogously to the torsion case, the generalized dual Hodge is defined in such a way to take into account all possible contractions,

$$\begin{aligned} \star R^{\alpha\beta}{}_{\mu\nu} = h\varepsilon_{\mu\nu\rho\sigma}\big[a R^{\alpha\beta\rho\sigma} + b\big(R^{\alpha\rho\beta\sigma} - R^{\beta\rho\alpha\sigma}\big) \\ + c\big(g^{\alpha\rho} R^{\beta\sigma} - g^{\beta\rho} R^{\alpha\sigma}\big) + d g^{\alpha\rho} g^{\beta\sigma} R\big], \end{aligned} \tag{8.29}$$

with a, b, c, d constant coefficients. By requiring, as in (8.17), that

$$\star\star R^{\alpha\beta}{}_{\mu\nu} = -R^{\alpha\beta}{}_{\mu\nu}, \tag{8.30}$$

we obtain a system of algebraic equations for a, b, c, d, which has the unique solution

$$a = 1/2 \quad \text{and} \quad b = c = d = 0. \tag{8.31}$$

For the curvature, therefore, the generalized dual coincides with the usual expression, that is,

$$\star R^{\alpha\beta}{}_{\mu\nu} = \frac{h}{2} \varepsilon_{\mu\nu\rho\sigma} R^{\alpha\beta\rho\sigma}. \tag{8.32}$$

This means that the additional index contractions related to soldering do not generate any additional contributions to the dual of curvature.

Comment 8.1 It is well known that *linear* General Relativity has duality symmetry [4]. That is to say, when written for the dual of curvature, the linearized Bianchi identity for curvature [see Eq. (1.89)] coincides with the linearized Einstein equation. Considering that Teleparallel Gravity is equivalent to General Relativity, it must also present duality symmetry in the linear case. In Chap. 15 we are going to show that, as a matter of fact, Teleparallel Gravity does have duality symmetry provided the dual torsion be written as given by Eq. (8.26). This shows, not only the consistency of the generalized dual definition for soldered bundles, but actually the necessity of such definition.

References

1. Frankel, T.: The Geometry of Physics. Cambridge University Press, Cambridge (1997)
2. Kobayashi, S., Nomizu, K.: Foundations of Differential Geometry, 2nd edn. Wiley-Interscience, New York (1996)
3. Gribl Lucas, T., Pereira, J.G.: J. Phys. A **42**, 035402 (2009). 0811.2066 [gr-qc]
4. Henneaux, M., Teitelboim, C.: Phys. Rev. D **71**, 024018 (2005). gr-qc/0408101

Chapter 9
Lagrangian and Field Equations

The lagrangian of Teleparallel Gravity, like those of gauge theories, is quadratic in the field strength—here represented by the torsion tensor. Up to a divergence, it is equivalent to the Einstein-Hilbert lagrangian of General Relativity. The teleparallel field equations, Bianchi identities and conservation laws are obtained. For the sake of comparison, the generalized teleparallel gravity with three free-parameters, also known as New General Relativity, is briefly discussed.

9.1 Lagrangian of Teleparallel Gravity

As a gauge theory for the translation group Teleparallel Gravity puts forward, as field strength, the torsion tensor

$$\overset{\bullet}{T} = \tfrac{1}{2} \overset{\bullet}{T}{}^a{}_{\mu\nu} P_a \, dx^\mu \wedge dx^\nu, \tag{9.1}$$

a 2-form assuming values in the Lie algebra of the gauge group. Like any gauge theory, the action of Teleparallel Gravity is of the form

$$\overset{\bullet}{\mathscr{S}} = \frac{c^3}{16\pi G} \int \mathrm{Tr}\big(\overset{\bullet}{T} \wedge \star \overset{\bullet}{T}\big), \tag{9.2}$$

with

$$\star \overset{\bullet}{T} = \tfrac{1}{2} \big(\star \overset{\bullet}{T}{}^a{}_{\rho\sigma}\big) P_a \, dx^\rho \wedge dx^\sigma \tag{9.3}$$

the corresponding dual form. The trace operation of Eq. (9.2) requires the presence of a metric on the group manifold. However, since the translation group is abelian, its Cartan-Killing bilinear form is degenerate and cannot be used. It is then necessary to look for another invariant metric. To find it out, we recall that the group manifold of translations is just the Minkowski spacetime $\mathbb{M}$, the quotient between the Poincaré ($\mathscr{P}$) and the Lorentz ($\mathscr{L}$) groups,

$$\mathbb{M} = \mathscr{P} / \mathscr{L}.$$

Although the Cartan-Killing metric does not exist in this case, the invariant Lorentz metric η_{ab} is also a metric on the group manifold $\mathbb{M}$, and can be used in its stead.

R. Aldrovandi, J.G. Pereira, *Teleparallel Gravity*, Fundamental Theories of Physics 173,
DOI 10.1007/978-94-007-5143-9_9, © Springer Science+Business Media Dordrecht 2013

Comment 9.1 This is quite similar to electromagnetism, a gauge theory for the abelian $U(1)$ group. Also in this case the Cartan-Killing metric must be replaced by a different invariant metric. In general the trivial one-dimensional metric $\eta = +1$ is chosen, though the choice $\eta = -1$ would also be possible. This second possibility will be discussed in more detail in Chap. 16, where the teleparallel equivalent of the Kaluza-Klein theory will be presented.

Action (9.2) can then be written as

$$\overset{\bullet}{\mathcal{S}} = \frac{c^3}{16\pi G} \int \eta_{ab} \overset{\bullet}{T}{}^a \wedge \star \overset{\bullet}{T}{}^b, \tag{9.4}$$

or, equivalently,

$$\overset{\bullet}{\mathcal{S}} = \frac{c^3}{64\pi G} \int \eta_{ab} \overset{\bullet}{T}{}^a{}_{\mu\nu} (\star \overset{\bullet}{T}{}^b{}_{\rho\sigma})\, dx^\mu \wedge dx^\nu \wedge dx^\rho \wedge dx^\sigma. \tag{9.5}$$

Taking into account that

$$dx^\mu \wedge dx^\nu \wedge dx^\rho \wedge dx^\sigma = -\varepsilon^{\mu\nu\rho\sigma} h\, d^4x, \tag{9.6}$$

with $h = \sqrt{-g}$, the action functional reduces to

$$\overset{\bullet}{\mathcal{S}} = -\frac{c^3}{64\pi G} \int \overset{\bullet}{T}{}_{\lambda\mu\nu} (\star \overset{\bullet}{T}{}^\lambda{}_{\rho\sigma}) \varepsilon^{\mu\nu\rho\sigma} h\, d^4x, \tag{9.7}$$

where we have identified

$$\overset{\bullet}{T}{}^\lambda{}_{\rho\sigma} = h_a{}^\lambda \overset{\bullet}{T}{}^a{}_{\rho\sigma}. \tag{9.8}$$

Using the generalized Hodge dual (8.26), as well as identity (1.92), the action assumes the form [1]

$$\overset{\bullet}{\mathcal{S}} = \frac{c^3}{32\pi G} \int \overset{\bullet}{T}{}_{\rho\mu\nu} \overset{\bullet}{S}{}^{\rho\mu\nu} h\, d^4x, \tag{9.9}$$

where

$$\overset{\bullet}{S}{}^{\rho\mu\nu} = -\overset{\bullet}{S}{}^{\rho\nu\mu} = \overset{\bullet}{K}{}^{\mu\nu\rho} - g^{\rho\nu} \overset{\bullet}{T}{}^{\sigma\mu}{}_\sigma + g^{\rho\mu} \overset{\bullet}{T}{}^{\sigma\nu}{}_\sigma \tag{9.10}$$

is the tensor introduced in Eq. (8.23), from now on called superpotential, and

$$\overset{\bullet}{K}{}^\nu{}_{\rho\mu} = \tfrac{1}{2} (\overset{\bullet}{T}{}_\rho{}^\nu{}_\mu + \overset{\bullet}{T}{}_\mu{}^\nu{}_\rho - \overset{\bullet}{T}{}^\nu{}_{\rho\mu}) \tag{9.11}$$

is the contortion tensor, with

$$\overset{\bullet}{T}{}^\nu{}_{\rho\mu} = \overset{\bullet}{\Gamma}{}^\nu{}_{\mu\rho} - \overset{\bullet}{\Gamma}{}^\nu{}_{\rho\mu} \tag{9.12}$$

the torsion of the Weitzenböck connection.

The lagrangian density corresponding to action (9.9) is

$$\overset{\bullet}{\mathcal{L}} = \frac{c^4 h}{32\pi G} \overset{\bullet}{T}{}_{\rho\mu\nu} \overset{\bullet}{S}{}^{\rho\mu\nu}. \tag{9.13}$$

Making use of the identity

$$\overset{\bullet}{T}{}^\mu{}_{\mu\rho} = \overset{\bullet}{K}{}^\mu{}_{\rho\mu}, \tag{9.14}$$

it can be rewritten as

$$\overset{\bullet}{\mathscr{L}} = \frac{c^4 h}{16\pi G}\big(\overset{\bullet}{K}{}^{\mu\nu\rho}\overset{\bullet}{K}{}_{\rho\nu\mu} - \overset{\bullet}{K}{}^{\mu\rho}{}_{\mu}\overset{\bullet}{K}{}^{\nu}{}_{\rho\nu}\big). \tag{9.15}$$

Substituting $\overset{\bullet}{K}{}^{\rho\mu\nu}$, we find

$$\overset{\bullet}{\mathscr{L}} = \frac{c^4 h}{16\pi G}\big(\tfrac{1}{4}\overset{\bullet}{T}{}^{\rho}{}_{\mu\nu}\overset{\bullet}{T}{}_{\rho}{}^{\mu\nu} + \tfrac{1}{2}\overset{\bullet}{T}{}^{\rho}{}_{\mu\nu}\overset{\bullet}{T}{}^{\nu\mu}{}_{\rho} - \overset{\bullet}{T}{}^{\rho}{}_{\mu\rho}\overset{\bullet}{T}{}^{\nu\mu}{}_{\nu}\big). \tag{9.16}$$

The first term corresponds to the usual lagrangian of internal gauge theories. The existence of the other two terms is related to the soldered character of the bundle. In fact, the presence of a tetrad field allows internal and external indices to be treated on the same footing, and consequently new contractions turn out to be possible—as discussed in Chap. 8. In terms of algebraic-indexed torsion, the teleparallel lagrangian assumes the form

$$\overset{\bullet}{\mathscr{L}} = \frac{c^4 h}{16\pi G}\big(\tfrac{1}{4}\overset{\bullet}{T}{}^{a}{}_{bc}\overset{\bullet}{T}{}_{a}{}^{bc} + \tfrac{1}{2}\overset{\bullet}{T}{}^{a}{}_{bc}\overset{\bullet}{T}{}^{cb}{}_{a} - \overset{\bullet}{T}{}^{a}{}_{ba}\overset{\bullet}{T}{}^{cb}{}_{c}\big). \tag{9.17}$$

Notice that torsion is a Lorentz tensor—it transforms covariantly under local Lorentz transformations. It then follows that each term of this lagrangian is local Lorentz invariant, and consequently the whole lagrangian is also invariant independently of the numerical value of the coefficients.

Comment 9.2 If we write a similar lagrangian with three arbitrary parameters α, β, γ, and with the coefficient of anholonomy replacing torsion, we get

$$\overset{\bullet}{\mathscr{L}} = \frac{c^4 h}{16\pi G}\big(\alpha f^{a}{}_{bc} f_{a}{}^{bc} + \beta f^{a}{}_{bc} f^{cb}{}_{a} + \gamma f^{a}{}_{ba} f^{cb}{}_{c}\big). \tag{9.18}$$

Since $f^{a}{}_{bc}$ is not a tensor [see Eq. (1.82)], this lagrangian is not—at least in principle—invariant under local Lorentz transformations. However, up to a surface term, it turns out to be invariant provided the constant parameters satisfy the relations [2]

$$\beta = 2\alpha \quad \text{and} \quad \gamma = -4\alpha. \tag{9.19}$$

For $\alpha = 1/4$, we get exactly the same coefficients of (9.17).

9.2 Equivalence with Einstein-Hilbert

As discussed in Chap. 4, the curvature of the Weitzenböck connection vanishes identically:

$$\overset{\bullet}{R}{}^{\rho}{}_{\lambda\nu\mu} \equiv \partial_\nu \overset{\bullet}{\Gamma}{}^{\rho}{}_{\lambda\mu} - \partial_\mu \overset{\bullet}{\Gamma}{}^{\rho}{}_{\lambda\nu} + \overset{\bullet}{\Gamma}{}^{\rho}{}_{\eta\nu}\overset{\bullet}{\Gamma}{}^{\eta}{}_{\lambda\mu} - \overset{\bullet}{\Gamma}{}^{\rho}{}_{\eta\mu}\overset{\bullet}{\Gamma}{}^{\eta}{}_{\lambda\nu} = 0. \tag{9.20}$$

Substituting the relation

$$\overset{\bullet}{\Gamma}{}^{\rho}{}_{\mu\nu} = \overset{\circ}{\Gamma}{}^{\rho}{}_{\mu\nu} + \overset{\bullet}{K}{}^{\rho}{}_{\mu\nu}, \tag{9.21}$$

we find

$$\overset{\bullet}{R}{}^{\rho}{}_{\theta\mu\nu} \equiv \overset{\circ}{R}{}^{\rho}{}_{\theta\mu\nu} + \overset{\bullet}{Q}{}^{\rho}{}_{\theta\mu\nu} = 0, \tag{9.22}$$

where

$$\overset{\circ}{R}{}^{\rho}{}_{\theta\mu\nu} = \partial_{\mu}\overset{\circ}{\Gamma}{}^{\rho}{}_{\theta\nu} - \partial_{\nu}\overset{\circ}{\Gamma}{}^{\rho}{}_{\theta\mu} + \overset{\circ}{\Gamma}{}^{\rho}{}_{\sigma\mu}\overset{\circ}{\Gamma}{}^{\sigma}{}_{\theta\nu} - \overset{\circ}{\Gamma}{}^{\rho}{}_{\sigma\nu}\overset{\circ}{\Gamma}{}^{\sigma}{}_{\theta\mu} \tag{9.23}$$

is the curvature of the Levi-Civita connection, and

$$\begin{aligned} \dot{Q}^{\rho}{}_{\theta\mu\nu} = {} & \partial_{\mu}\dot{K}^{\rho}{}_{\theta\nu} - \partial_{\nu}\dot{K}^{\rho}{}_{\theta\mu} + \dot{\Gamma}^{\rho}{}_{\sigma\mu}\dot{K}^{\sigma}{}_{\theta\nu} - \dot{\Gamma}^{\rho}{}_{\sigma\nu}\dot{K}^{\sigma}{}_{\theta\mu} \\ & - \dot{\Gamma}^{\sigma}{}_{\theta\mu}\dot{K}^{\rho}{}_{\sigma\nu} + \dot{\Gamma}^{\sigma}{}_{\theta\nu}\dot{K}^{\rho}{}_{\sigma\mu} + \dot{K}^{\rho}{}_{\sigma\nu}\dot{K}^{\sigma}{}_{\theta\mu} - \dot{K}^{\rho}{}_{\sigma\mu}\dot{K}^{\sigma}{}_{\theta\nu} \end{aligned} \tag{9.24}$$

is a tensor written in terms of the Weitzenböck connection only. Like the Riemann curvature tensor, it is a 2-form assuming values in the Lie algebra of the Lorentz group:

$$\dot{Q} = \tfrac{1}{2} S_a{}^b \dot{Q}^a{}_{b\mu\nu}\, dx^{\mu} \wedge dx^{\nu}. \tag{9.25}$$

Its components are given by

$$\dot{Q}^a{}_{b\mu\nu} = \dot{\mathscr{D}}_{\mu}\dot{K}^a{}_{b\nu} - \dot{\mathscr{D}}_{\nu}\dot{K}^a{}_{b\mu} + \dot{K}^a{}_{c\nu}\dot{K}^c{}_{b\mu} - \dot{K}^a{}_{c\mu}\dot{K}^c{}_{b\nu}, \tag{9.26}$$

with

$$\dot{\mathscr{D}}_{\mu}\dot{K}^a{}_{b\nu} = \partial_{\mu}\dot{K}^a{}_{b\nu} + \dot{A}^a{}_{c\mu}\dot{K}^c{}_{b\nu} - \dot{A}^c{}_{b\mu}\dot{K}^a{}_{c\nu} \tag{9.27}$$

the Fock-Ivanenko derivative of the contortion 1-form $\dot{K}^a{}_{b\nu}$.

Identity (9.22) reads

$$-\overset{\circ}{R}{}^{\rho}{}_{\theta\mu\nu} = \dot{Q}^{\rho}{}_{\theta\mu\nu}. \tag{9.28}$$

By taking appropriate contractions, the scalar version of this equation is found to be

$$-\overset{\circ}{R} = \dot{Q} \equiv \left(\dot{K}^{\mu\nu\rho}\dot{K}_{\rho\nu\mu} - \dot{K}^{\mu\rho}{}_{\mu}\dot{K}^{\nu}{}_{\rho\nu}\right) + \frac{2}{h}\partial_{\mu}\left(h\dot{T}^{\nu\mu}{}_{\nu}\right), \tag{9.29}$$

with $\overset{\circ}{R}$ the scalar curvature of the Levi-Civita connection. Comparing with the teleparallel lagrangian (9.15), we see that

$$\dot{\mathscr{L}} = \overset{\circ}{\mathscr{L}} - \partial_{\mu}\left(\frac{c^4 h}{8\pi G}\dot{T}^{\nu\mu}{}_{\nu}\right), \tag{9.30}$$

where

$$\overset{\circ}{\mathscr{L}} = -\frac{c^4}{16\pi G}\sqrt{-g}\,\overset{\circ}{R} \tag{9.31}$$

is the Einstein-Hilbert lagrangian of General Relativity. Up to a divergence, therefore, the lagrangian of Teleparallel Gravity is equivalent to the lagrangian of General Relativity.

To understand the reason for the presence of a divergence term between the two lagrangians, let us recall [see Sect. 3.2] that the Einstein-Hilbert lagrangian (9.31) depends on the metric, as well as on the first and second derivatives of the metric. Equivalently, in the context of the tetrad formalism, we can say that it depends on the tetrad, as well as on the first and second derivatives of the tetrad field. The terms containing second derivatives, however, reduce to a divergence term [3]. In consequence, it is possible to rewrite the Einstein-Hilbert lagrangian in a form stating this aspect explicitly:

$$\overset{\circ}{\mathscr{L}} = \overset{\circ}{\mathscr{L}}_1 + \partial_\mu\big(\sqrt{-g}\,w^\mu\big), \tag{9.32}$$

where $\overset{\circ}{\mathscr{L}}_1$ is a lagrangian that depends solely on the tetrad and on its first derivatives, and w^μ is a four-vector. On the other hand, the teleparallel lagrangian $\overset{\bullet}{\mathscr{L}}$ depends only on the tetrad and on its first derivative. The divergence in the equivalence relation (9.30) is then necessary to remove the terms containing second derivatives of the tetrad from the Einstein-Hilbert lagrangian $\overset{\circ}{\mathscr{L}}$.

As is well-known, it is not possible to construct an invariant lagrangian for General Relativity in terms of the tetrad and its first derivatives only. This means that $\overset{\circ}{\mathscr{L}}_1$ is not by itself invariant. However, since it changes by a pure divergence, there are actually infinitely many different such first-order lagrangians $\overset{\circ}{\mathscr{L}}_1$, each one with its own particular surface term $\sqrt{-g}\,w^\mu$. One specific first-order lagrangian for General Relativity is the Møller lagrangian [4]

$$\overset{\circ}{\mathscr{L}}_{\text{M}} = \frac{c^4 h}{16\pi G}\big(\overset{\circ}{\nabla}_\mu h^{a\nu}\overset{\circ}{\nabla}_\nu h_a{}^\mu - \overset{\circ}{\nabla}_\mu h_a{}^\mu\overset{\circ}{\nabla}_\nu h^{a\nu}\big). \tag{9.33}$$

It is invariant under diffeomorphisms, but not under local Lorentz transformations. The interesting point of this lagrangian is that, in the class of frames h'_a in which the teleparallel spin connection vanishes, this lagrangian is found to coincide exactly—that is, without any surface term—with the teleparallel lagrangian:

$$\overset{\bullet}{\mathscr{L}} = \overset{\circ}{\mathscr{L}}_{\text{M}}. \tag{9.34}$$

Of course, in any other class of frames a surface term will emerge in the equivalence relation.

9.3 Matter Energy-Momentum Density

Let us denote by

$$\mathscr{S}_s = \frac{1}{c}\int \mathscr{L}_s\, d^4x \tag{9.35}$$

the action integral of a general source field. The lagrangian $\mathscr{L}_s$ is assumed to depend on the source fields and on their first derivatives only. When coupled to gravitation, it turns out to depend also on the tetrad and on its first derivatives. Under an arbitrary variation $\delta h_a{}^\mu$ of the tetrad field, the action variation is then written in the form

$$\delta\mathscr{S}_s = \frac{1}{c}\int \Theta^a{}_\mu \delta h_a{}^\mu h\, d^4x, \tag{9.36}$$

where

$$h\Theta^a{}_\mu = \frac{\delta\mathscr{L}_s}{\delta h_a{}^\mu} \equiv \frac{\partial\mathscr{L}_s}{\partial h_a{}^\mu} - \partial_\rho\frac{\partial\mathscr{L}_s}{\partial_\rho\partial h_a{}^\mu} \tag{9.37}$$

is the matter energy-momentum tensor. Since the tetrad is linear in the translational gauge potential $B^a{}_\mu$,

$$h^a{}_\mu = \dot{\mathscr{D}}_\mu x^a + B^a{}_\mu, \tag{9.38}$$

the matter energy-momentum tensor can equally be written in the form

$$h\Theta^a{}_\mu = \frac{\delta \mathscr{L}_s}{\delta B_a{}^\mu} \equiv \frac{\partial \mathscr{L}_s}{\partial B_a{}^\mu} - \partial_\rho \frac{\partial \mathscr{L}_s}{\partial_\rho \partial B_a{}^\mu}, \tag{9.39}$$

which is consistent with the gauge structure of Teleparallel Gravity.

Let us consider first an infinitesimal local (point-dependent) Lorentz transformation

$$\Lambda_a{}^b = \delta_a^b + \varepsilon_a{}^b,$$

with $\varepsilon_a{}^b$ $(= -\varepsilon^b{}_a)$ the transformation parameters. Under such a transformation, the tetrad changes according to

$$\delta h_a{}^\mu = \varepsilon_a{}^b h_b{}^\mu. \tag{9.40}$$

Substituting in the action variation (9.36), we obtain

$$\delta \mathscr{S}_s = \frac{1}{c} \int \Theta^a{}_b \varepsilon_a{}^b h \, d^4x. \tag{9.41}$$

Since $\varepsilon_a{}^b$ is anti-symmetric, the requirement of invariance of the action under local Lorentz transformations implies that the energy-momentum tensor

$$\Theta^a{}_b = \Theta^a{}_\mu h_b{}^\mu \tag{9.42}$$

must be symmetric:

$$\Theta^a{}_b = \Theta_b{}^a. \tag{9.43}$$

In this case, as is well known, spin and angular momentum are separately conserved. This is the basic output of the local Lorentz invariance of the matter action [5].

Let us consider now a general transformation of the spacetime coordinates,

$$x'^\rho = x^\rho + \xi^\rho. \tag{9.44}$$

Under such a transformation, the tetrad changes according to

$$\delta h_a{}^\mu \equiv h'_a{}^\mu(x) - h_a{}^\mu(x) = h_a{}^\rho \partial_\rho \xi^\mu - \xi^\rho \partial_\rho h_a{}^\mu. \tag{9.45}$$

Substituting in the action variation (9.36), we obtain

$$\delta \mathscr{S}_s = \frac{1}{c} \int \Theta^a{}_\mu \left[h_a{}^\rho \partial_\rho \xi^\mu - \xi^\rho \partial_\rho h_a{}^\mu \right] h \, d^4x \tag{9.46}$$

or, equivalently,

$$\delta \mathscr{S}_s = \frac{1}{c} \int \left[\Theta^\rho{}_c \partial_\rho \xi^c - \xi^c \Theta^\rho{}_\mu \partial_\rho h_c{}^\mu - \Theta^a{}_\mu \xi^\rho \partial_\rho h_a{}^\mu \right] h \, d^4x, \tag{9.47}$$

where $\xi^c = h^c{}_\mu \xi^\mu$. Using the identity

$$\partial_\rho h_a{}^\mu = \dot{A}^b{}_{a\rho} h_b{}^\mu - \dot{\Gamma}^\mu{}_{\lambda\rho} h_a{}^\lambda, \tag{9.48}$$

which follows from Eq. (4.66), and making use of the symmetry of the energy-momentum tensor, the action variation assumes the form

$$\delta \mathscr{S}_s = \frac{1}{c} \int \Theta^\rho{}_c \big[\partial_\rho \xi^c + \big(\overset{\bullet}{A}{}^c{}_{b\rho} - \overset{\bullet}{K}{}^c{}_{b\rho}\big)\xi^b\big] h\, d^4x. \tag{9.49}$$

Integrating the first term by parts and neglecting the surface term, the invariance of the action yields

$$\int \big[\partial_\mu \big(h\Theta_a{}^\mu\big) - \big(\overset{\bullet}{A}{}^b{}_{a\mu} - \overset{\bullet}{K}{}^b{}_{a\mu}\big)\big(h\Theta_b{}^\mu\big)\big]\xi^a\, d^4x = 0. \tag{9.50}$$

From the arbitrariness of ξ^a, it follows that

$$\overset{\bullet\bullet}{\mathscr{D}}_\mu \big(h\Theta_a{}^\mu\big) \equiv \partial_\mu \big(h\Theta_a{}^\mu\big) - \big(\overset{\bullet}{A}{}^b{}_{a\mu} - \overset{\bullet}{K}{}^b{}_{a\mu}\big)\big(h\Theta_b{}^\mu\big) = 0. \tag{9.51}$$

Making use of the identity

$$\partial_\rho h = h \overset{\circ}{\Gamma}{}^\nu{}_{\nu\rho} \equiv h\big(\overset{\bullet}{\Gamma}{}^\nu{}_{\rho\nu} - \overset{\bullet}{K}{}^\nu{}_{\rho\nu}\big), \tag{9.52}$$

the above conservation law becomes

$$\partial_\mu \Theta_a{}^\mu + \big(\overset{\bullet}{\Gamma}{}^\mu{}_{\rho\mu} - \overset{\bullet}{K}{}^\mu{}_{\rho\mu}\big)\Theta_a{}^\rho - \big(\overset{\bullet}{A}{}^b{}_{a\mu} - \overset{\bullet}{K}{}^b{}_{a\mu}\big)\Theta_b{}^\mu = 0. \tag{9.53}$$

In a purely spacetime form, it reads

$$\overset{\bullet\bullet}{\nabla}_\mu \Theta_\lambda{}^\mu \equiv \partial_\mu \Theta_\lambda{}^\mu + \big(\overset{\bullet}{\Gamma}{}^\mu{}_{\rho\mu} - \overset{\bullet}{K}{}^\mu{}_{\rho\mu}\big)\Theta_\lambda{}^\rho - \big(\overset{\bullet}{\Gamma}{}^\rho{}_{\lambda\mu} - \overset{\bullet}{K}{}^\rho{}_{\lambda\mu}\big)\Theta_\rho{}^\mu = 0. \tag{9.54}$$

This is the conservation law of the source energy-momentum tensor in Teleparallel Gravity. Of course, due to the relation (9.21), it coincides with the corresponding conservation law of General Relativity:

$$\overset{\circ}{\nabla}_\mu \Theta_\lambda{}^\mu \equiv \partial_\mu \Theta_\lambda{}^\mu + \overset{\circ}{\Gamma}{}^\mu{}_{\rho\mu}\Theta_\lambda{}^\rho - \overset{\circ}{\Gamma}{}^\rho{}_{\lambda\mu}\Theta_\rho{}^\mu = 0. \tag{9.55}$$

This is actually a matter of consistency as the conservation law must be unique.

Comment 9.3 "Covariant conservation laws" are not, strictly speaking, true conservation laws because they fail to yield a conserved "charge"—that is, a time-conserved quantity. They are actually identities, called Noether identities, which rule the exchange of energy and momentum between the source and the gravitational field [6].

9.4 Field Equations

Consider now the lagrangian

$$\mathscr{L} = \overset{\bullet}{\mathscr{L}} + \mathscr{L}_s, \tag{9.56}$$

with $\mathscr{L}_s$ the lagrangian of a general source (matter) field. Introducing the notation

$$k = \frac{8\pi G}{c^4}, \tag{9.57}$$

variation with respect to the gauge potential $B^a{}_\rho$—or equivalently, with respect to the tetrad field $h^a{}_\mu$—yields the teleparallel version of the gravitational field equation [7]

$$\partial_\sigma\big(h\overset{\bullet}{S}_a{}^{\rho\sigma}\big) - kh\overset{\bullet}{J}_a{}^\rho = kh\Theta_a{}^\rho. \tag{9.58}$$

In this equation,

$$h\overset{\bullet}{S}_a{}^{\rho\sigma} = -k\frac{\partial\overset{\bullet}{\mathcal{L}}}{\partial(\partial_\sigma h^a{}_\rho)} \tag{9.59}$$

represents the superpotential, whereas the term

$$h\overset{\bullet}{J}_a{}^\rho = -\frac{\partial\overset{\bullet}{\mathcal{L}}}{\partial h^a{}_\rho} \tag{9.60}$$

stands for the gauge current, which in this case represents the Noether energy-momentum density of gravitation itself [8]. Finally,

$$h\Theta_a{}^\rho = -\frac{\delta\mathcal{L}_s}{\delta h^a{}_\rho} \equiv -\left(\frac{\partial\mathcal{L}_s}{\partial h^a{}_\rho} - \partial_\mu\frac{\partial\mathcal{L}_s}{\partial_\mu\partial h^a{}_\rho}\right) \tag{9.61}$$

is the matter (or source) energy-momentum tensor. In Appendix C we present a detailed computation leading to the results

$$\overset{\bullet}{S}_a{}^{\rho\sigma} = \overset{\bullet}{K}{}^{\rho\sigma}{}_a - h_a{}^\sigma\overset{\bullet}{T}{}^{\theta\rho}{}_\theta + h_a{}^\rho\overset{\bullet}{T}{}^{\theta\sigma}{}_\theta \tag{9.62}$$

and

$$\overset{\bullet}{J}_a{}^\rho = \frac{1}{k}h_a{}^\mu\overset{\bullet}{S}_c{}^{\nu\rho}\overset{\bullet}{T}{}^c{}_{\nu\mu} - \frac{h_a{}^\rho}{h}\overset{\bullet}{\mathcal{L}} + \frac{1}{k}\overset{\bullet}{A}{}^c{}_{a\sigma}\overset{\bullet}{S}_c{}^{\rho\sigma}. \tag{9.63}$$

Due to the anti-symmetry of the superpotential in the last two indices, the total—that is, gravitational plus source—energy-momentum density is conserved in the ordinary sense:

$$\partial_\rho\big(h\overset{\bullet}{J}_a{}^\rho + h\Theta_a{}^\rho\big) = 0. \tag{9.64}$$

The left-hand side of the gravitational field equation (9.58) depends on torsion only. Using the identity (9.21), through a lengthy but straightforward calculation, it can be shown to satisfy

$$\partial_\sigma\big(h\overset{\bullet}{S}_a{}^{\rho\sigma}\big) - k\big(h\overset{\bullet}{J}_a{}^\rho\big) = h\big(\overset{\circ}{R}_a{}^\rho - \tfrac{1}{2}h_a{}^\rho\overset{\circ}{R}\big). \tag{9.65}$$

We see from this expression that, as expected due to the equivalence between the corresponding lagrangians, the teleparallel field equation (9.58) is equivalent to Einstein's field equation

$$\overset{\circ}{R}_a{}^\rho - \tfrac{1}{2}h_a{}^\rho\overset{\circ}{R} = k\Theta_a{}^\rho. \tag{9.66}$$

Observe that the energy-momentum tensor appears as the source in both theories: as the source of curvature in General Relativity, and as the source of torsion in Teleparallel Gravity. This shows once more that, according to Teleparallel Gravity, curvature and torsion are related to the same degrees of freedom of the gravitational field.

Comment 9.4 There are gravitational models in which curvature and torsion are related to different degrees of freedom of gravity. One of such models is the so-called New General Relativity, which will be briefly discussed in Sect. 9.6. Another example is the Einstein-Cartan model, which will be discussed in Chap. 17.

9.5 Bianchi Identities

The Bianchi identities of Teleparallel Gravity [9] can be obtained from the general Bianchi identities [presented in Sect. 1.5] by replacing

$$A^a{}_{b\mu} \to \dot{A}^a{}_{b\mu} - \dot{K}^a{}_{b\mu}, \tag{9.67}$$

which implies the concomitant replacement

$$\mathscr{D}_\mu = \partial_\mu - iA_\mu \quad \to \quad \ddot{\mathscr{D}}_\mu = \partial_\mu - i\big(\dot{A}^a{}_{b\mu} - \dot{K}^a{}_{b\mu}\big), \tag{9.68}$$

with $\ddot{\mathscr{D}}_\mu = \mathring{\mathscr{D}}_\mu$ the teleparallel equivalent of the General Relativity Fock-Ivanenko covariant derivative. Similarly, torsion and curvature must be replaced according to

$$T^a{}_{\mu\nu} \to \dot{T}^a{}_{\mu\nu} - \dot{T}^a{}_{\mu\nu} = 0 \tag{9.69}$$

and

$$R^a{}_{b\mu\nu} \to -\dot{Q}^a{}_{b\mu\nu} = \mathring{R}^a{}_{b\mu\nu}. \tag{9.70}$$

Hence the Bianchi identity for torsion, which in the general case is given by [see Eq. (1.84)]

$$\mathscr{D}_\nu T^a{}_{\rho\mu} + \mathscr{D}_\mu T^a{}_{\nu\rho} + \mathscr{D}_\rho T^a{}_{\mu\nu} = R^a{}_{\rho\mu\nu} + R^a{}_{\nu\rho\mu} + R^a{}_{\mu\nu\rho}, \tag{9.71}$$

assumes the teleparallel form

$$\dot{Q}^\rho{}_{\theta\mu\nu} + \dot{Q}^\rho{}_{\nu\theta\mu} + \dot{Q}^\rho{}_{\mu\nu\theta} = 0, \tag{9.72}$$

which is clearly equivalent to the first Bianchi identity of General Relativity

$$\mathring{R}^\rho{}_{\theta\mu\nu} + \mathring{R}^\rho{}_{\nu\theta\mu} + \mathring{R}^\rho{}_{\mu\nu\theta} = 0. \tag{9.73}$$

Through a tedious but straightforward calculation, Bianchi identity (9.72) can be rewritten as

$$\dot{\mathscr{D}}_\nu \dot{T}^a{}_{\rho\mu} + \dot{\mathscr{D}}_\mu \dot{T}^a{}_{\nu\rho} + \dot{\mathscr{D}}_\rho \dot{T}^a{}_{\mu\nu} = 0. \tag{9.74}$$

Comment 9.5 In the class of frames h'_a in which $\dot{A}'^a{}_{b\mu} = 0$, the Bianchi identity (9.74) reduces to

$$\partial_\nu \dot{T}'^a{}_{\rho\mu} + \partial_\mu \dot{T}'^a{}_{\nu\rho} + \partial_\rho \dot{T}'^a{}_{\mu\nu} = 0. \tag{9.75}$$

In this form it becomes quite similar to the Bianchi identity of electromagnetism,

$$\partial_\nu F_{\rho\mu} + \partial_\mu F_{\nu\rho} + \partial_\rho F_{\mu\nu} = 0, \tag{9.76}$$

with $F_{\mu\nu}$ the electromagnetic field strength.

On the other hand, the Bianchi identity for curvature, which in the general case is given by [see Eq. (1.85)]

$$\mathscr{D}_\nu R^a{}_{b\rho\mu} + \mathscr{D}_\mu R^a{}_{b\nu\rho} + \mathscr{D}_\rho R^a{}_{b\mu\nu} = 0, \tag{9.77}$$

acquires in Teleparallel Gravity the form

$$\ddot{\mathscr{D}}_\nu \dot{Q}^a{}_{b\rho\mu} + \ddot{\mathscr{D}}_\mu \dot{Q}^a{}_{b\nu\rho} + \ddot{\mathscr{D}}_\rho \dot{Q}^a{}_{b\mu\nu} = 0, \tag{9.78}$$

which is clearly equivalent to the General Relativity Bianchi identity

$$\mathring{\mathscr{D}}_\nu \mathring{R}^a{}_{b\rho\mu} + \mathring{\mathscr{D}}_\mu \mathring{R}^a{}_{b\nu\rho} + \mathring{\mathscr{D}}_\rho \mathring{R}^a{}_{b\mu\nu} = 0. \tag{9.79}$$

As is well known, the contracted form of this identity is

$$\mathring{\mathscr{D}}_\rho \big[h \big(\mathring{R}_a{}^\rho - \tfrac{1}{2} h_a{}^\rho \mathring{R} \big) \big] = 0. \tag{9.80}$$

Through a similar procedure, the contracted form of the teleparallel Bianchi identity (9.78) is found to be

$$\ddot{\mathscr{D}}_\rho \big[\partial_\sigma \big(h \dot{S}_a{}^{\rho\sigma} \big) - k h \dot{J}_a{}^\rho \big] = 0. \tag{9.81}$$

Recalling that, in the presence of a source field, the teleparallel field equation is

$$\partial_\sigma \big(h \dot{S}_a{}^{\rho\sigma} \big) - k h \dot{J}_a{}^\rho = k h \Theta_a{}^\rho, \tag{9.82}$$

the Bianchi identity (9.81) is seen to be consistent with the conservation law

$$\ddot{\mathscr{D}}_\rho \big(h \Theta_a{}^\rho \big) = 0, \tag{9.83}$$

as obtained from Noether's theorem [see Sect. 9.3].

Comment 9.6 It is worth mentioning that the Bianchi identity for curvature (9.78) has no equivalent in internal (Yang-Mills type) gauge theories. This may come as a surprise because, as we have repeatedly said, the bundle of internal gauge theories is not soldered, and consequently torsion cannot be defined. Notwithstanding, it is the Bianchi identity for torsion that has a counterpart in internal gauge theories. The Bianchi identity for curvature does not have it. In Chap. 16, where the teleparallel equivalent of the Kaluza-Klein model is studied, this point will resurface: in this model, the electromagnetic field strength, which represents the curvature of the internal space, appears as additional components of torsion, not of curvature.

9.6 A Glimpse into New General Relativity

Like in Teleparallel Gravity, the spin connection of New General Relativity [10] is the purely inertial connection

$$\dot{A}^b{}_{c\mu} = \Lambda^b{}_d(x) \partial_\mu \Lambda_c{}^d(x) \tag{9.84}$$

introduced in Chap. 2. As already discussed, it is a connection with vanishing curvature and, for a non-trivial tetrad, non-vanishing torsion:

$$\dot{R}^a{}_{b\mu\nu} = 0 \quad \text{and} \quad \dot{T}^a{}_{\mu\nu} \neq 0. \tag{9.85}$$

The spacetime-indexed linear connection corresponding to the inertial spin connection (9.84) is the Weitzenböck connection

$$\dot{\Gamma}^{\rho}{}_{\nu\mu} = h_a{}^{\rho}\partial_{\mu}h^a{}_{\nu} + h_a{}^{\rho}\dot{A}^a{}_{b\mu}h^b{}_{\nu} \equiv h_a{}^{\rho}\dot{\mathscr{D}}_{\mu}h^a{}_{\nu}. \tag{9.86}$$

The corresponding torsion tensor is

$$\dot{T}^{\rho}{}_{\mu\nu} = \dot{\Gamma}^{\rho}{}_{\nu\mu} - \dot{\Gamma}^{\rho}{}_{\mu\nu}. \tag{9.87}$$

New General Relativity is a generalized teleparallel model with three arbitrary parameters a_1, a_2, a_3:

$$\dot{\mathscr{L}}_{ngr} = \frac{c^4 h}{16\pi G}\left(a_1 \dot{T}^{\rho}{}_{\mu\nu}\dot{T}_{\rho}{}^{\mu\nu} + a_2 \dot{T}^{\rho}{}_{\mu\nu}\dot{T}^{\nu\mu}{}_{\rho} + a_3 \dot{T}^{\rho}{}_{\mu\rho}\dot{T}^{\nu\mu}{}_{\nu}\right). \tag{9.88}$$

In terms of irreducible components of torsion [see Sect. 1.7], it can be rewritten in the form,

$$\dot{\mathscr{L}}_{ngr} = \frac{c^4 h}{16\pi G}\left(b_1 \dot{\mathscr{T}}^{\rho}{}_{\mu\nu}\dot{\mathscr{T}}_{\rho}{}^{\mu\nu} + b_2 \dot{\mathscr{V}}^{\mu}\dot{\mathscr{V}}_{\mu} + b_3 \dot{\mathscr{A}}^{\mu}\dot{\mathscr{A}}_{\mu}\right), \tag{9.89}$$

with b_1, b_2, b_3 new constant coefficients. Considering that, up to a divergence,

$$-\tfrac{2}{3}\dot{\mathscr{T}}^{\rho}{}_{\mu\nu}\dot{\mathscr{T}}_{\rho}{}^{\mu\nu} + \tfrac{2}{3}\dot{\mathscr{V}}^{\mu}\dot{\mathscr{V}}_{\mu} - \tfrac{3}{2}\dot{\mathscr{A}}^{\mu}\dot{\mathscr{A}}_{\mu} = -\overset{\circ}{R}, \tag{9.90}$$

with $\overset{\circ}{R}$ the scalar curvature of the Levi-Civita connection, lagrangian (9.89) can be recast as

$$\dot{\mathscr{L}}_{ngr} = \frac{c^4 h}{16\pi G}\left(-\overset{\circ}{R} + c_1 \dot{\mathscr{T}}^{\rho}{}_{\mu\nu}\dot{\mathscr{T}}_{\rho}{}^{\mu\nu} + c_2 \dot{\mathscr{V}}^{\mu}\dot{\mathscr{V}}_{\mu} + c_3 \dot{\mathscr{A}}^{\mu}\dot{\mathscr{A}}_{\mu}\right), \tag{9.91}$$

with the new coefficients given by

$$c_1 = b_1 + \tfrac{2}{3}, \qquad c_2 = b_2 - \tfrac{2}{3}, \qquad c_3 = b_3 + \tfrac{3}{2}. \tag{9.92}$$

The first term of (9.91) is the Einstein-Hilbert lagrangian. In this theory, therefore, torsion is assumed to produce deviations from the predictions of General Relativity—or equivalently, from the predictions of Teleparallel Gravity. This means that in this theory torsion represents additional degrees of freedom in relation to curvature.

In New General Relativity, any matter field Ψ is assumed to be minimally coupled to the teleparallel spin connection [10]:

$$\partial_{\mu}\Psi \to \dot{\mathscr{D}}_{\mu}\Psi = \partial_{\mu}\Psi - \frac{i}{2}\dot{A}^{ab}{}_{\mu}S_{ab}\Psi. \tag{9.93}$$

However, as we have seen in Chap. 5, the strong equivalence principle says that the gravitational coupling prescription is minimal only in the spin connection of General Relativity. This means that the coupling prescription (9.93) violates the principle. Furthermore, since the teleparallel spin connection vanishes in a specific class of frames, the above prescription implies actually that any matter field should couple trivially to gravitation,

$$\partial_{\mu}\Psi \to \dot{\mathscr{D}}_{\mu}\Psi = \partial_{\mu}\Psi, \tag{9.94}$$

which is of course a physically unacceptable result. It should be remarked also that solar system experiments severely restrict the existence of non-vanishing c_1 and c_2 in the lagrangian (9.91). Furthermore, it has already been shown that the Schwarzschild solution exists only for the case with [11]

$$c_1 = c_2 = c_3 = 0.$$

In principle, therefore, and taking into account the compelling experimental evidences for the existence of black holes in the universe, one can say that New General Relativity lacks experimental support.

References

1. Maluf, J.W.: J. Math. Phys. **35**, 335 (1994)
2. Cho, Y.M.: Phys. Rev. D **14**, 2521 (1976)
3. Landau, L.D., Lifshitz, E.M.: The Classical Theory of Fields. Pergamon, Oxford (1975)
4. Møller, C.: K. Dan. Vidensk. Selsk. Mat. Fys. Skr. **89**, 13 (1978)
5. Weinberg, S.: Gravitation and Cosmology. Wiley, New York (1972)
6. Konopleva, N.P., Popov, V.N.: Gauge Fields. Harwood, New York (1980)
7. Hehl, F.W.: Four lectures on Poincaré gauge field theory. In: Bergmann, P.G., de Sabbata, V. (eds.) Cosmology and Gravitation: Spin, Torsion, Rotation, and Supergravity, p. 52. Plenum, New York (1980)
8. de Andrade, V.C., Guillen, L.C.T., Pereira, J.G.: Phys. Rev. Lett. **84**, 4533 (2000). gr-qc/0003100
9. de Andrade, V.C., Guillen, L.C.T., Pereira, J.G.: Teleparallel gravity: an overview. In: Gurzadyan, V.G., Jantzen, R.T., Ruffini, R. (eds.) Proceedings of the Ninth Marcel Grossmann Meeting. World Scientific, Singapore (2002). gr-qc/0011087
10. Hayashi, K., Shirafuji, T.: Phys. Rev. D **19**, 3524 (1979)
11. Obukhov, Yu.N., Pereira, J.G.: Phys. Rev. D **67**, 044008 (2003). gr-qc/0212078

Chapter 10
Gravitational Energy-Momentum Density

In Teleparallel Gravity it is possible to obtain separate expressions for the energy-momentum density of gravity and of the interaction of gravity with inertial effects of the frame. The energy-momentum density of gravity shows up as a true tensor, and satisfies a covariant conservation law. The energy-momentum density associated to the inertial effects is neither conserved nor covariant. Together, they form a pseudotensor conserved in the ordinary sense. This means that the non-covariance of the usual expressions for the gravitational energy-momentum density is not an intrinsic property of gravity, but a consequence of the fact that they include also the energy-momentum density related to the inertial effects.

10.1 Introduction

It is natural for many physicists to expect that, as for any fundamental field, gravitation have its own *local* energy-momentum density. For example, in the preface to his classic textbook [1], Synge says that *in Einstein's theory, either there is a gravitational field or there is none, according to as the Riemann tensor does not or does vanish. This is an absolute property*; *it has nothing to do with any observer's world line*. In a similar stroke, Bondi [2] argues that *in relativity a non-localizable form of energy is inadmissible, because any form of energy contributes to gravitation and so its location can in principle be found*. According to these arguments, the energy of the gravitational field should be localizable independently of the observer.

On the other hand, owing perhaps to the inherent difficulty to find its explicit form in the context of General Relativity, it is usually accepted that such a density cannot be locally defined because of the equivalence principle [3]. According to this point of view, any attempt to identify an energy-momentum density for the gravitational field leads to complexes that are not true tensors. The first of such attempts was made by Einstein himself, who proposed an expression which was nothing but the canonical expression obtained from Noether's theorem [4]. This quantity is, like many others, a pseudotensor, an object that depends on the coordinate system. Several other attempts have been made, leading to different expressions for this pseudotensor.

R. Aldrovandi, J.G. Pereira, *Teleparallel Gravity*, Fundamental Theories of Physics 173, DOI 10.1007/978-94-007-5143-9_10,

Despite the existence of some controversial points related to the equivalence principle [1, 5], it seems true that in the context of General Relativity no tensorial expression for the gravitational energy-momentum density exists. In the stream of this perception, a *quasilocal* approach [6] has been proposed which, although it does not solve the problem, sheds nevertheless some light on it [7, 8]. According to this approach, to each gravitational energy-momentum pseudotensor is associated a *superpotential* which is a hamiltonian boundary term. The energy-momentum defined by such a pseudotensor does not really depend on the local reference frame, but only on its values on the boundary of a region—from which its *quasilocal* character. As the relevant boundary conditions are physically acceptable, this approach is said to validate the pseudotensor approach to the gravitational energy-momentum problem. Independently of this and others attempts to circumvent the problem of the gravitational energy-momentum density, the question remains: is the impossibility of defining a tensorial expression for the gravitational energy-momentum density a fundamental property of gravity, or just a consequence of the particular geometrical description of General Relativity?

10.2 Field Equations and Conservation Laws

In the first-order formalism, where the lagrangian depends on the tetrad (or metric) and on its first derivatives only, the gravitational field equation can be obtained from the usual (first-order) Euler-Lagrange equation

$$\frac{\partial \mathscr{L}}{\partial h^a{}_\rho} - \partial_\sigma \frac{\partial \mathscr{L}}{\partial(\partial_\sigma h^a{}_\rho)} = 0. \tag{10.1}$$

This equation can be rewritten in the form

$$\partial_\sigma \left(h S_a{}^{\rho\sigma} \right) - k h J_a{}^\rho = 0, \tag{10.2}$$

where $k = 8\pi G/c^4$,

$$S_a{}^{\rho\sigma} = -S_a{}^{\sigma\rho} \equiv -\frac{k}{h} \frac{\partial \mathscr{L}}{\partial(\partial_\sigma h^a{}_\rho)} \tag{10.3}$$

is the superpotential, and

$$J_a{}^\rho \equiv -\frac{1}{h} \frac{\partial \mathscr{L}}{\partial h^a{}_\rho} \tag{10.4}$$

stands for the Noether gravitational energy-momentum current. Equation (10.2) is known as the *potential form* of the gravitational field equation [9]. It is, in structure, similar to the Yang-Mills equation. Its main virtue is to explicitly exhibit the complex defining the Noether energy-momentum current of the gravitational field. In fact, due to the anti-symmetry of $S_a{}^{\rho\sigma}$ in the last two indices, the field equation implies the conservation of the gravitational energy-momentum current:

$$\partial_\rho \left(h J_a{}^\rho \right) = 0. \tag{10.5}$$

Any conservation law written as a four-dimensional *ordinary* divergence is a *true conservation law* in the sense that it yields a time-conserved "charge". The procedure is quite simple. Integrate the conservation law (10.5) in a space section of spacetime, and separate the partial derivative into time and space components:

$$\int \partial_0 \big(h J_a{}^0\big)\, d^3x = -\int \partial_k \big(h J_a{}^k\big)\, d^3x. \tag{10.6}$$

Using Gauss theorem, the integral on the right-hand side can be transformed into a surface integral at infinity,

$$\int \partial_0 \big(h J_a{}^0\big)\, d^3x = -\oint \mathbf{J}_a \cdot d\boldsymbol{\sigma}. \tag{10.7}$$

Supposing that all physical fields must vanish at infinity, the right-hand side vanishes identically. Moving the time derivative to outside the integration sign, we then get

$$\frac{dq_a}{dt} = 0, \tag{10.8}$$

with

$$q_a = \int J_a{}^0 h\, d^3x \tag{10.9}$$

the time-conserved charge.

On the other hand, in order to be physically meaningful, the equation expressing any conservation law must be covariant under both general coordinate and local Lorentz transformations. Although trivial in the absence of gravitation, this simple property has an important consequence in the gravitational case: *no tensorial quantity can be truly conserved.* In fact, since the derivative in a true conservation law is not covariant, in order to get a covariant conservation law, the conserved current cannot be covariant either. This means that the Noether energy-momentum current $J_a{}^\rho$ appearing in the gravitational field equation (10.2), which has the ordinary conservation law (10.5), cannot be a tensor.

Conversely, since the symmetric energy-momentum complex $\Theta_a{}^\rho$ of a matter field—the object appearing as source in the right-hand side of the gravitational field equation—is a true tensor, it can be conserved in the covariant sense only,

$$\overset{\circ}{\mathscr{D}}_\rho \big(h \Theta_a{}^\rho\big) \equiv h \big(\partial_\rho \Theta_a{}^\rho - \overset{\circ}{A}{}^c{}_{a\rho} \Theta_c{}^\rho + \overset{\circ}{\Gamma}{}^\rho{}_{\lambda\rho} \Theta_a{}^\lambda\big) = 0, \tag{10.10}$$

otherwise the conservation law itself would not be covariant. We reinforce that this kind of "conservation law" is not a true conservation law, but just an identity (called Noether identity) regulating the exchange of energy-momentum between matter and gravitation. We can then conclude that, if a *tensorial* expression $t_a{}^\rho$ for the energy-momentum density of gravitation exists, it must necessarily be conserved in the covariant sense.

10.3 Teleparallel Gravity

As seen in Chap. 9, the (sourceless) gravitational field equation of Teleparallel Gravity is

$$\partial_\sigma \big(h \overset{\bullet}{S}_a{}^{\rho\sigma} \big) - k h \overset{\bullet}{J}_a{}^{\rho} = 0, \tag{10.11}$$

where

$$\overset{\bullet}{S}_a{}^{\rho\sigma} \equiv -\frac{k}{h} \frac{\partial \overset{\bullet}{\mathscr{L}}}{\partial(\partial_\sigma h^a{}_\rho)} = \overset{\bullet}{K}{}^{\rho\sigma}{}_a - h_a{}^{\sigma} \overset{\bullet}{T}{}^{\theta\rho}{}_\theta + h_a{}^{\rho} \overset{\bullet}{T}{}^{\theta\sigma}{}_\theta \tag{10.12}$$

is the superpotential, and

$$\overset{\bullet}{J}_a{}^{\rho} \equiv -\frac{1}{h} \frac{\partial \overset{\bullet}{\mathscr{L}}}{\partial h^a{}_\rho} = \frac{1}{k} h_a{}^{\lambda} \overset{\bullet}{S}_c{}^{\nu\rho} \overset{\bullet}{T}{}^{c}{}_{\nu\lambda} - \frac{h_a{}^{\rho}}{h} \overset{\bullet}{\mathscr{L}} + \frac{1}{k} \overset{\bullet}{A}{}^{c}{}_{a\sigma} \overset{\bullet}{S}_c{}^{\rho\sigma} \tag{10.13}$$

is the energy-momentum current.

In absence of gravitation, the anholonomy of the frames is entirely related to the inertial forces present in those frames. The preferred class of inertial frames is characterized by the absence of such inertial effects. In the presence of gravitation, on the other hand, the anholonomy of the frames is related to *both* gravitational and inertial effects. This means that there are actually no holonomic frames in the presence of gravitation. Like in Special Relativity, however, also in the presence of gravitation there is a preferred class of frames: those which reduce to the inertial class in the absence of gravitation. Since in Teleparallel Gravity inertial effects are represented by the spin connection, this preferred class of frames, which we denote by h'_b, is characterized by the (global) vanishing of the teleparallel spin connection:

$$\overset{\bullet}{A}{}'^{a}{}_{b\mu} = 0. \tag{10.14}$$

Comment 10.1 It is important to remark that, as discussed in Sect. 6.5, this has nothing to do with the strong equivalence principle, which says that the Levi-Civita connection of General Relativity can be made to vanish at a point, or along a trajectory.

In a Lorentz rotated frame $h_a = \Lambda_a{}^b h'_b$, that connection assumes the form

$$\overset{\bullet}{A}{}^{a}{}_{b\mu} = \Lambda^a{}_e \partial_\mu \Lambda_b{}^e. \tag{10.15}$$

This is the spin connection of Teleparallel Gravity. As in Special Relativity, it represents solely the inertial properties of the frame, not gravitation.

We are now approaching the main point: put together, the last term of the current (10.13) and the potential term of field equation (10.11) make up a Fock-Ivanenko covariant derivative (which, we recall, acts only on the algebraic indices) of the superpotential:

$$\partial_\sigma \big(h \overset{\bullet}{S}_a{}^{\rho\sigma} \big) - \overset{\bullet}{A}{}^{c}{}_{a\sigma} \big(h \overset{\bullet}{S}_c{}^{\rho\sigma} \big) \equiv \overset{\bullet}{\mathscr{D}}_\sigma \big(h \overset{\bullet}{S}_a{}^{\rho\sigma} \big). \tag{10.16}$$

This allows us to rewrite that field equation in the form

$$\overset{\bullet}{\mathscr{D}}_\sigma \big(h \overset{\bullet}{S}_a{}^{\rho\sigma} \big) - k h \overset{\bullet}{t}_a{}^{\rho} = 0, \tag{10.17}$$

where

$$\overset{\bullet}{t}_a{}^\rho = \frac{1}{k} h_a{}^\lambda \overset{\bullet}{S}_c{}^{\nu\rho} \overset{\bullet}{T}{}^c{}_{\nu\lambda} - \frac{h_a{}^\rho}{h} \overset{\bullet}{\mathscr{L}} \tag{10.18}$$

is a tensorial current. The crucial point comes out finally: since the teleparallel spin connection (10.15) has vanishing curvature, the corresponding Fock-Ivanenko derivative turns out to be commutative:

$$[\overset{\bullet}{\mathscr{D}}_\rho, \overset{\bullet}{\mathscr{D}}_\sigma] = 0. \tag{10.19}$$

Taking into account the anti-symmetry of the superpotential in the last two indices, it follows from (10.17) that the tensorial current (10.18) is covariantly conserved:

$$\overset{\bullet}{\mathscr{D}}_\rho (h \overset{\bullet}{t}_a{}^\rho) = 0. \tag{10.20}$$

Due to these properties, $\overset{\bullet}{t}_a{}^\rho$ can be interpreted as the energy-momentum density of gravitation alone. Accordingly, the last term in (10.13),

$$\overset{\bullet}{\imath}_a{}^\rho = \frac{1}{k} \overset{\bullet}{A}{}^c{}_{a\sigma} \overset{\bullet}{S}_c{}^{\rho\sigma}, \tag{10.21}$$

can be interpreted as the energy-momentum density associated to the coupling of gravitation with the inertial effects of the frame. The total energy-momentum density is consequently

$$\overset{\bullet}{\jmath}_a{}^\rho = \overset{\bullet}{t}_a{}^\rho + \overset{\bullet}{\imath}_a{}^\rho, \tag{10.22}$$

which is non-covariant because of the inertial effects. Due to the anti-symmetry of the superpotential in the last two indices, we see from the field equation (10.11) that this current is conserved in the ordinary sense:

$$\partial_\rho (h \overset{\bullet}{\jmath}_a{}^\rho) = 0. \tag{10.23}$$

This is actually a matter of consistency: since the current is not covariant, the derivative appearing in the conservation law cannot be covariant either, otherwise the conservation law itself would not be covariant.

The reason for the usual expressions for the gravitational energy-momentum density to be a pseudotensor, therefore, is that they include, in addition to the energy-momentum density of gravitation, a contribution coming from the interaction of gravitation with the inertial effects of the frame. When considered separately from inertia, the gravitational energy-momentum density is found to be a true tensor and conserved in the covariant sense. As discussed in Sect. 10.2, this is consistent with the property that a covariant quantity can only be conserved in the covariant sense, otherwise the conservation law itself would be physically meaningless. The energy-momentum density associated with the inertia effects, on the other hand, is neither conserved nor covariant. Nevertheless, once put together with the energy-momentum tensor of gravity, it does constitute a pseudotensor conserved in the ordinary sense. We can then conclude that the non-covariance of the usual general-relativistic expressions for the gravitational energy-momentum density is not an intrinsic property of gravity, but a consequence of the fact that they include also the energy-momentum density associated to the inertial effects of the frame, which is non-tensorial by its very nature.

Comment 10.2 Owing to its odd asymptotic behavior, the contribution of the inertial effects often yields unphysical (divergent or trivial) results for the total energy and momentum of a gravitational system. As a consequence, it is necessary to make use of a regularizing process in the computation in order to eliminate the spurious contribution coming from those inertial effects [10]. The existence of a purely gravitational energy-momentum tensor in Teleparallel Gravity allows one to compute unequivocally the energy and momentum of any gravitational system without necessity of a regularization process [11].

An interesting property of the tensorial current (10.18) comes out if we remember the expression (9.13) of the teleparallel lagrangian: as field theory would expect for a massless field, its trace vanishes identically:

$$\overset{\bullet}{t}_\rho{}^\rho \equiv h^a{}_\rho \overset{\bullet}{t}_a{}^\rho = 0. \tag{10.24}$$

On the other hand, the trace of the total pseudocurrent (10.13) is found to be proportional to the very lagrangian itself:

$$h\overset{\bullet}{j}_\rho{}^\rho \equiv hh^a{}_\rho \overset{\bullet}{j}_a{}^\rho = -\frac{h}{2k} \overset{\bullet}{S}_a{}^{\mu\nu} \overset{\bullet}{T}{}^a{}_{\mu\nu} = -2\overset{\bullet}{\mathcal{L}}. \tag{10.25}$$

Similar results hold for the symmetric and the canonical energy-momentum densities of the electromagnetic field, given respectively by [12]

$$\Theta_\lambda{}^\rho = -F_\lambda{}^\mu F^\rho{}_\mu + \tfrac{1}{4}\delta_\lambda{}^\rho F_{\mu\nu} F^{\mu\nu} \tag{10.26}$$

and

$$\theta_\lambda{}^\rho = -\partial_\lambda A^\mu F^\rho{}_\mu + \tfrac{1}{4}\delta_\lambda{}^\rho F_{\mu\nu} F^{\mu\nu}. \tag{10.27}$$

In fact, as a simple calculation shows,

$$\Theta_\rho{}^\rho = 0 \quad \text{and} \quad \theta_\rho{}^\rho = -2\mathcal{L}_{em}.$$

One can then say that $\overset{\bullet}{t}_a{}^\rho$ and $\overset{\bullet}{j}_a{}^\rho$ play roles similar to those of the symmetric and the canonical energy-momentum tensors of massless source fields. This similarity becomes still more evident if we note that, whereas the symmetric energy-momentum tensor $\Theta_\lambda{}^\rho$ of the electromagnetic field is gauge invariant, the canonical tensor $\theta_\lambda{}^\rho$ is not, as it depends explicitly on the (partial derivative of the) electromagnetic potential A^μ.

10.4 General Relativity

The question then arises: would it be possible to define a purely gravitational energy-momentum *tensor* in the context of General Relativity? To begin with, we recall that it is impossible to construct an *invariant* lagrangian for General Relativity in terms of the tetrad and its first derivatives only. What exists is the second-order Einstein-Hilbert invariant lagrangian, in which the second-derivative terms reduce to a total divergence. This lagrangian is of the form

$$\overset{\circ}{\mathcal{L}} = -\frac{h}{2k} \overset{\circ}{R} \equiv \overset{\circ}{\mathcal{L}}_1 + \partial_\mu(hw^\mu), \tag{10.28}$$

where $\overset{\circ}{\mathscr{L}}_1$ is a first-order lagrangian and w^μ is a contravariant four-vector. As mentioned at the end of Sect. 9.2, there are actually infinitely many different first-order lagrangians, each one connected to a different surface term:

$$\overset{\circ}{\mathscr{L}} = \overset{\circ}{\mathscr{L}}_1 + \partial_\mu\big(hw^\mu\big) = \overset{\circ}{\mathscr{L}}{}'_1 + \partial_\mu\big(hw'^\mu\big) = \overset{\circ}{\mathscr{L}}{}''_1 + \partial_\mu\big(hw''^\mu\big) = \cdots. \tag{10.29}$$

Considering that the divergence term does not contribute to the field equation, any one of the first-order lagrangians, when substituted in the Euler-Lagrange equation

$$\frac{\partial \overset{\circ}{\mathscr{L}}_1}{\partial h^a{}_\rho} - \partial_\sigma \frac{\partial \overset{\circ}{\mathscr{L}}_1}{\partial(\partial_\sigma h^a{}_\rho)} = 0, \tag{10.30}$$

will lead to the potential form of Einstein equation,

$$\partial_\sigma\big(h\overset{\circ}{S}_a{}^{\rho\sigma}\big) - kh\overset{\circ}{J}_a{}^\rho = 0, \tag{10.31}$$

with superpotential

$$\overset{\circ}{S}_a{}^{\rho\sigma} = -\frac{k}{h}\frac{\partial \overset{\circ}{\mathscr{L}}_1}{\partial(\partial_\sigma h^a{}_\rho)} \tag{10.32}$$

and energy-momentum current

$$\overset{\circ}{J}_a{}^\rho = -\frac{1}{h}\frac{\partial \overset{\circ}{\mathscr{L}}_1}{\partial h^a{}_\rho}. \tag{10.33}$$

Due to the anti-symmetry of the superpotential in the last two indices, this current is conserved in the ordinary sense:

$$\partial_\rho\big(h\overset{\circ}{J}_a{}^\rho\big) = 0. \tag{10.34}$$

As a consequence, it is necessarily a pseudotensor. Since there are infinitely many first-order lagrangians, it is possible to define infinitely many pseudo-currents, each one connected to a different superpotential. Some examples can be found in the papers [13–22].

Let us now follow the same steps of the teleparallel case studied in the previous section, and rewrite Einstein's equation (10.31) in the form

$$\overset{\circ}{\mathscr{D}}_\sigma\big(h\overset{\circ}{S}_a{}^{\rho\sigma}\big) - kh\overset{\circ}{t}_a{}^\rho = 0, \tag{10.35}$$

where

$$\overset{\circ}{\mathscr{D}}_\sigma\big(h\overset{\circ}{S}_a{}^{\rho\sigma}\big) = \partial_\sigma\big(h\overset{\circ}{S}_a{}^{\rho\sigma}\big) - \overset{\circ}{A}{}^b{}_{a\sigma}\big(h\overset{\circ}{S}_b{}^{\rho\sigma}\big) \tag{10.36}$$

is the covariant derivative in the General Relativity spin connection $\overset{\circ}{A}{}^b{}_{a\sigma}$, and

$$h\overset{\circ}{t}_a{}^\rho = h\overset{\circ}{J}_a{}^\rho - \frac{h}{k}\overset{\circ}{A}{}^b{}_{a\sigma}\overset{\circ}{S}_b{}^{\rho\sigma} \tag{10.37}$$

is a tensorial quantity that, at least in principle, could be interpreted as the energy-momentum density of gravitation.

Comment 10.3 That $h\overset{\circ}{t}_a{}^\rho$ is a covariant quantity can be verified by observing that, since Einstein's equation (10.35) is covariant, and considering that its first term is also covariant, the current term must necessarily be a tensor.

There is, however, a problem with this interpretation: unlike the derivative $\overset{\bullet}{\mathscr{D}}_\sigma$ of the teleparallel case, the covariant derivative $\overset{\circ}{\mathscr{D}}_\sigma$ appearing in the field equation (10.35) is not commutative. Instead of complying with a property of type (10.19), it actually satisfies

$$[\overset{\circ}{\mathscr{D}}_\sigma, \overset{\circ}{\mathscr{D}}_\lambda](h\overset{\circ}{S}_a{}^{\rho\sigma}) = \overset{\circ}{R}{}^b{}_{a\sigma\lambda}(h\overset{\circ}{S}_b{}^{\rho\sigma}). \tag{10.38}$$

It follows that the tensorial current (10.37) is not covariantly conserved,

$$\overset{\circ}{\mathscr{D}}_\rho(h\overset{\circ}{t}_a{}^\rho) \neq 0, \tag{10.39}$$

and for this reason it is not physically meaningful. In General Relativity, therefore, it is not possible to define a tensorial expression for the gravitational energy-momentum density. The basic reason is that inertial and gravitational effects are both embodied in the spin connection $\overset{\circ}{A}{}^a{}_{b\mu}$, and cannot be separated. Because of this inseparability, the energy-momentum current in General Relativity always include, in addition to the purely gravitational density, also the energy-momentum density associated with inertial effects of the frame. Since the latter is a pseudotensor, the energy-momentum complex will be always a pseudotensor.

Comment 10.4 Conversely, we could say that the requirement of covariant conservation of $\overset{\circ}{t}_a{}^\rho$ would impose unphysical constraints on the spacetime geometry. This problem is similar to the inconsistencies that appear in the theory of a fundamental spin-2 field coupled to gravitation in the context of general relativity [23, 24]. Considering that Teleparallel Gravity is able to give a consistent answer to the problem of the localizability of the gravitational energy-momentum density, one would wonder whether it could also shed some light on those inconsistencies. This question will be discussed in more detail in Chap. 13.

It is worth mentioning here an interesting proposal made by T. Levi-Civita soon after General Relativity was introduced [25]. According to him, the whole left-hand side of Einstein equation

$$\overset{\circ}{R}_a{}^\rho - \tfrac{1}{2}h_a{}^\rho\overset{\circ}{R} = k\Theta_a{}^\rho \tag{10.40}$$

should be interpreted as (minus) the gravitational energy-momentum density. As an immediate consequence, the total (source plus gravitational) energy-momentum density of any gravitational system should be zero. If we write Einstein equation in the potential form,

$$\partial_\sigma(h\overset{\circ}{S}_a{}^{\rho\sigma}) - kh\overset{\circ}{J}_a{}^\rho = kh\Theta_a{}^\rho, \tag{10.41}$$

this proposal acquires an additional appeal. It is well-known that, in field theory, the canonical energy-momentum of a field with non-vanishing spin is not necessarily symmetric [see, for example, Eq. (10.27)]. A symmetrization procedure is necessary to make it acceptable as a source in Einstein's equation. The most commonly used is the Belinfante-Rosenfeld method [26, 27], in which a symmetric tensor is obtained by adding a total derivative of a convenient anti-symmetric tensor. In the above equation, the left-hand side can be interpreted as a kind of Belinfante-Rosenfeld energy-momentum tensor for spin-endowed fields, with the superpotential term—the divergence of an anti-symmetric tensor—playing just the role of the additional symmetrizing term.

10.5 Comparison with the Gauge Self-current

The problem of defining an energy-momentum density for the gravitational field has a close analogy to the problem of defining a gauge self-current for the Yang-Mills field, discussed in Sect. 3.1. The sourceless Yang-Mills equation is [28]

$$\partial_\mu F^{A\mu\nu} - j^{A\nu} = 0, \tag{10.42}$$

with $F^{A\mu\nu}$ the curvature of the connection $A^A{}_\mu$. The piece

$$j^{A\nu} = -f^A{}_{BC} A^B{}_\mu F^{C\mu\nu} \tag{10.43}$$

represents the gauge self-current. Owing to the anti-symmetry of $F^{A\mu\nu}$ in the last two indices, this current is conserved:

$$\partial_\nu j^{A\nu} = 0. \tag{10.44}$$

However, due to the explicit presence of the connection $A^B{}_\mu$ in its expression, this current is clearly not gauge covariant, in analogy with the gravitational energy-momentum pseudotensor. There is a difference, though: whereas the gravitational energy-momentum pseudotensor is made up of a tensorial plus a non-tensorial part [see Eq. (10.22)], the Yang-Mills self-current consists of a gauge non-covariant part only. In fact, considering that $F^{A\mu\nu}$ belongs to the adjoint representation, its covariant derivative has the form

$$D_\mu F^{A\mu\nu} = \partial_\mu F^{A\mu\nu} + f^A{}_{BC} A^B{}_\mu F^{C\mu\nu}, \tag{10.45}$$

which allows the sourceless Yang-Mills equation to be rewritten in the form

$$D_\mu F^{A\mu\nu} = 0. \tag{10.46}$$

On the other hand, the gravitational field equation reads

$$\overset{\bullet}{\mathscr{D}}_\sigma \left(h \overset{\bullet}{S}_a{}^{\rho\sigma} \right) - k h \overset{\bullet}{t}_a{}^\rho = 0, \tag{10.47}$$

with $\overset{\bullet}{t}_a{}^\rho$ the tensorial part of the gravitational current. A comparison between Eqs. (10.46) and (10.47) shows that, differently from the gravitational case, the gauge self-current does not have a covariant piece. These properties are consistent with the idea that the Yang-Mills field (or gluons in Quantum Chromodynamics) cannot propagate freely in space, whereas gravitational waves—at least in principle—are allowed to exist freely in spacetime.

Comment 10.5 In the specific case of Electromagnetism, whose sourceless field equation is

$$\partial_\mu F^{\mu\nu} = 0, \tag{10.48}$$

the gauge self-current vanishes identically—it has neither a gauge covariant nor a gauge non-covariant part. Although they do not transport a gauge charge, however, electromagnetic waves are able to carry energy-momentum, and consequently to propagate freely in spacetime.

References

1. Synge, J.L.: Relativity: The General Theory. Wiley, New York (1960)
2. Bondi, H.: Proc. R. Soc. Lond. Ser. A **427**, 249 (1990)
3. Misner, C.W., Thorne, K.S., Wheeler, J.A.: Gravitation. Freeman, San Francisco (1973)
4. Trautman, A.: In: Witten, L. (ed.) Gravitation: An Introduction to Current Research. Wiley, New York (1962)
5. Damour, T.: Questioning the equivalence principle. C. R. Acad. Sci., Sér. IV **2**, 1249 (2001). gr-qc/0109063 (Contribution to the workshop Missions spatiales en physique fondamentale, Chatillon, France, 18–19 January 2001, edited by C. Bordé and P. Touboul)
6. Brown, J.D., York, J.W.: Phys. Rev. D **47**, 1407 (1993). gr-qc/9209012
7. Chang, C.C., Nester, J.M.: Phys. Rev. Lett. **83**, 1897 (1999). gr-qc/9809040
8. Szabados, L.B.: Living Rev. Relativ. **7**, 4 (2004)
9. Møller, C.: K. Dan. Vidensk. Selsk. Mat. Fys. Skr. **89**, 13 (1978)
10. Maluf, J.W., Veiga, M.V.O., da Rocha-Neto, J.F.: Gen. Relativ. Gravit. **39**, 227 (2007). gr-qc/0507122
11. Gribl Lucas, T., Obukhov, Yu.N., Pereira, J.G.: Phys. Rev. D **80**, 064043 (2009). 0909.2418 [gr-qc]
12. Landau, L.D., Lifshitz, E.M.: The Classical Theory of Fields. Pergamon, Oxford (1975)
13. Papapetrou, A.: Proc. R. Ir. Acad. A **52**, 11 (1948)
14. Bergmann, P.G., Thompson, R.: Phys. Rev. **89**, 400 (1953)
15. Møller, C.: Ann. Phys. **4**, 347 (1958)
16. Dubois-Violette, M., Madore, J.: Commun. Math. Phys. **108**, 213 (1987)
17. Szabados, L.B.: Class. Quantum Gravity **9**, 2521 (1992)
18. Aguirregabiria, J.M., Chamorro, A., Virbhadra, K.S.: Gen. Relativ. Gravit. **28**, 1393 (1996)
19. Maluf, J.W.: Gen. Relativ. Gravit. **30**, 413 (1998)
20. Deser, S., Franklin, J.S., Seminara, D.: Class. Quantum Gravity **16**, 2815 (1999)
21. Babak, S.V., Grishchuk, L.P.: Phys. Rev. D **61**, 024038 (2000)
22. Itin, Y.: Class. Quantum Gravity **19**, 173 (2002)
23. Aragone, C., Deser, S.: Nuovo Cimento A **3**, 4 (1971)
24. Aragone, C., Deser, S.: Nuovo Cimento B **57**, 1 (1980)
25. Levi-Civita, T.: Rend. R. Accad. Lincei **26**, 381 (1917). For an English translation (with a foreword), see Antocci, S., Loinger, A.: physics/9906004
26. Belinfante, F.J.: Physica **6**, 687 (1939)
27. Rosenfeld, L.: Mém. Acad. R. Belg. Sci. **18**, 1 (1940)
28. Itzykson, C., Zuber, J.B.: Quantum Field Theory. McGraw-Hill, New York (1980)

Chapter 11
Gravitation in the Lack of Universality

General Relativity is fundamentally grounded on the universality of the gravitational interaction, which is the basic statement of the weak equivalence principle. If universality somehow fails, the geometric description of General Relativity breaks down. Due to its gauge structure, Teleparallel Gravity does not require the weak equivalence principle, and remains a consistent theory in the lack of universality.

11.1 Gravitation and Universality

Universality of gravitation means that everything feels gravity the same. Provided the initial conditions are the same, all particles, independently of their masses and constitutions, will follow the same trajectory. In the study of a massive particle motion, universality of free fall is directly connected with the equality between inertial and gravitational masses: $m_i = m_g$. In fact, in order to be eliminated from the classical (newtonian) equation of motion—such that the motion becomes universal—they must necessarily coincide. It is important to remark that, even though newtonian gravity was constructed to comply with universality, it remains a consistent theory for $m_i \neq m_g$.

Einstein's General Relativity, on the other hand, is a theory fundamentally based on the universality of free fall or, equivalently, on the weak equivalence principle. In this theory, geometry replaces the concept of force in the description of the gravitational interaction, and there is no room for violation of universality. Despite its success in all experimental tests at the classical level [1], a possible violation of the weak equivalence principle would lead, among other consequences, to the non-universality of free fall, and consequently to its conceptual breakdown. Of course, like newtonian Gravity, it could still be used in most practical cases but, again like Newton's theory, it could fail in describing some specific physical situation.

Gravitation is the only universal interaction in Nature. It is, consequently, the only one to allow a description in terms of the geometry of spacetime, as is done by General Relativity. The non-universal character of electromagnetism is the reason why there is not such a geometric description for the electromagnetic interaction.

R. Aldrovandi, J.G. Pereira, *Teleparallel Gravity*, Fundamental Theories of Physics 173,
DOI 10.1007/978-94-007-5143-9_11, © Springer Science+Business Media Dordrecht 2013

What does exist, as is well known, is a gauge description. On the other hand, as a gauge theory for the translation group, the teleparallel equivalent of General Relativity does not describe the gravitational interaction through a geometrization of spacetime, but as a gravitational force similar to the Lorentz force of electrodynamics [see Sect. 6.2]. Since Maxwell theory, a gauge theory for the unitary group $U(1)$, is able to describe the non-universal electromagnetic interaction, the question then arises whether the gauge approach of Teleparallel Gravity would also be able to describe the gravitational interaction in the lack of universality, that is, in the absence of the weak equivalence principle. This is the issue we are going to address in this chapter.

11.2 The Electromagnetic Coupling Prescription

The non-universal electromagnetic interaction is described by a gauge theory for the unitary group $U(1)$. An element of this group is written as

$$U = \exp[i J \varepsilon], \tag{11.1}$$

where $\varepsilon = \varepsilon(x^\mu)$ is the transformation parameter, and J is the generator of infinitesimal transformations—a member of the Lie algebra. When applied to a general field Ψ representing a particle of electric charge q, the generator in the Ψ representation is written as [2]

$$J = \sqrt{\alpha_e} \equiv \sqrt{q^2/\hbar c}, \tag{11.2}$$

with α_e the *electromagnetic fine-structure constant* (in gaussian units). The gauge transformation of a general complex field Ψ,

$$\Psi' = U \Psi, \tag{11.3}$$

is then written as

$$\Psi' = \exp[i \sqrt{\alpha_e} \varepsilon] \Psi, \tag{11.4}$$

whose infinitesimal form is

$$\delta \Psi = i \sqrt{\alpha_e} \varepsilon \Psi. \tag{11.5}$$

Then comes the point: in order to render the derivative

$$D_\mu \Psi = (\partial_\mu + i \sqrt{\alpha_e} A_\mu) \Psi \tag{11.6}$$

really covariant, the gauge potential must transform according to

$$A'_\mu = A_\mu - \partial_\mu \varepsilon. \tag{11.7}$$

The electromagnetic coupling prescription is then given by

$$\partial_\mu \to \partial_\mu + i \sqrt{\alpha_e} A_\mu. \tag{11.8}$$

It is interesting to notice that in terms of the Planck charge

$$q_P = \sqrt{\hbar c}, \tag{11.9}$$

the square root of the fine-structure constant assumes the form

$$\sqrt{\alpha_e} = \frac{q}{q_P}. \tag{11.10}$$

For $q = q_P$, the fine-structure constant reduces to $\sqrt{\alpha_e} = 1$. We see from this expression that the intensity of the electromagnetic interaction depends on how different is the particle electric charge in relation to the Planck charge.

11.3 Gravitation Without Universality

Let us imagine now the existence of particles with gravitational and inertial masses which are different: $m_g \neq m_i$. Analogously to the electromagnetic case, whose gauge transformation involves the electromagnetic fine structure constant, the translational gauge transformation in the non-universal case must depend on some kind of *dimensionless coupling constant* α_u, which gives a measurement of the violation of universality.

11.3.1 Non-universal Coupling Prescription

To begin with, let us observe that the square root of the usual dimensionless gravitational constant for a particle of mass m is [3]

$$\sqrt{\alpha_G} = \sqrt{\frac{Gm_g^2}{\hbar c}} \equiv \frac{m_g}{m_P}, \tag{11.11}$$

with m_P the Planck mass. We see from this expression that the intensity of the gravitational interaction depends on how different is the particle gravitational mass m_g in relation to the Planck mass. This is quite similar to the square root of the fine-structure constant (11.10). Analogously, if one is interested in studying the effects coming from a violation of universality, the ensuing dimensionless coupling constant must depend on how different is the gravitational mass m_g in relation to the inertial mass m_i:

$$\sqrt{\alpha_u} = \frac{m_g}{m_i}. \tag{11.12}$$

In the universal case it reduces to $\alpha_u = 1$.

If Ψ denotes a field representing a particle with $m_g \neq m_i$. Its translational gauge transformation is

$$\Psi' = \tilde{U}\Psi, \tag{11.13}$$

with

$$\tilde{U} = \exp\big(\sqrt{\alpha_u}\varepsilon^a \partial_a\big) \tag{11.14}$$

an element of the translation group. Notice that, like in the electromagnetic case, it depends now on the particle properties. The corresponding infinitesimal transformation is

$$\tilde{\delta}\Psi = \sqrt{\alpha_u}\varepsilon^a \partial_a \Psi. \tag{11.15}$$

If we rewrite it the form

$$\tilde{\delta}\Psi = \tilde{\delta}x^a \partial_a \Psi, \tag{11.16}$$

we see that the non-universal gauge transformation of the tangent space coordinates is given by

$$\tilde{\delta}x^a = \sqrt{\alpha_u}\varepsilon^a. \tag{11.17}$$

Using now the general definition (4.32) of covariant derivative, the translational gauge covariant derivative of Ψ is found to be

$$\tilde{h}_\mu \Psi = \partial_\mu \Psi + \sqrt{\alpha_u} B^a{}_\mu \partial_a \Psi. \tag{11.18}$$

Similarly to the universal case, it can be rewritten in the form

$$\tilde{h}_\mu \Psi = \tilde{h}^a{}_\mu \partial_a \Psi, \tag{11.19}$$

where

$$\tilde{h}^a{}_\mu \equiv \tilde{h}_\mu x^a = \partial_\mu x^a + \sqrt{\alpha_u} B^a{}_\mu \tag{11.20}$$

is the translational covariant derivative of x^a. In a general Lorentz-rotated frame [see Eq. (2.7) and comments around it], it assumes the form

$$\tilde{h}^a{}_\mu = \partial_\mu x^a + \dot{A}^a{}_{b\mu} x^b + \sqrt{\alpha_u} B^a{}_\mu \tag{11.21}$$

or, equivalently,

$$\tilde{h}^a{}_\mu = \dot{\mathscr{D}}_\mu x^a + \sqrt{\alpha_u} B^a{}_\mu. \tag{11.22}$$

The non-universal gravitational coupling prescription is achieved by replacing

$$e^a{}_\mu = \dot{\mathscr{D}}_\mu x^a \quad \to \quad \tilde{h}^a{}_\mu = \dot{\mathscr{D}}_\mu x^a + \sqrt{\alpha_u} B^a{}_\mu. \tag{11.23}$$

The translational covariant derivative is then written as

$$\tilde{h}_\mu \Psi = \left(\partial_\mu x^a + \dot{A}^a{}_{b\mu} x^b + \sqrt{\alpha_u} B^a{}_\mu\right) \partial_a \Psi. \tag{11.24}$$

Comment 11.1 Observe that the breakdown of universality modifies the coupling of the particle to gravitation—represented by the translational gauge potential $B^a{}_\mu$—but does not change its coupling to the inertial effects of the frame—represented by the inertial connection $\dot{A}^a{}_{b\mu}$.

To see that the derivative (11.24) is in fact covariant, let us write its gauge-transformed version as

$$\tilde{h}'_\mu \Psi' = \dot{\mathscr{D}}_\mu x'^a \partial_a \Psi' + \sqrt{\alpha_u} B'^a{}_\mu \partial_a \Psi', \tag{11.25}$$

where we have used the gauge invariance of the generators ∂_a. Substituting

$$\Psi' = \Psi + \sqrt{\alpha_u}\varepsilon^c \partial_c \Psi, \tag{11.26}$$

as well as the gauge potential transformation

$$B'^a{}_\mu = B^a{}_\mu - \overset{\bullet}{\mathscr{D}}_\mu \varepsilon^a, \tag{11.27}$$

we see that

$$\tilde{\delta}(\tilde{h}_\mu \Psi) = \sqrt{\alpha_u}\varepsilon^a \partial_a(\tilde{h}_\mu \Psi), \tag{11.28}$$

which shows that $\tilde{h}_\mu \Psi$ is gauge covariant.

11.3.2 Particle Equation of Motion

If the inertial and gravitational masses of a given particle are different, its free action in a general frame is written as

$$\mathscr{S} = -m_i c \int_{\mathsf{p}}^{\mathsf{q}} u_a \overset{\bullet}{\mathscr{D}}_\mu x^a \, dx^\mu. \tag{11.29}$$

The corresponding action in the presence of gravitation is obtained by applying the coupling prescription (11.23), in which case we get

$$\mathscr{S} = -m_i c \int_{\mathsf{p}}^{\mathsf{q}} u_a \big(\overset{\bullet}{\mathscr{D}}_\mu x^a + \sqrt{\alpha_u} B^a{}_\mu\big) \, dx^\mu. \tag{11.30}$$

Using the gauge transformations (11.17) and (11.27), we see immediately that this action is gauge invariant. Furthermore, since the gauge potential is a Lorentz vector in the algebraic index, that is,

$$B'^a{}_\mu = \Lambda^a{}_b B^b{}_\mu, \tag{11.31}$$

the action is also locally Lorentz-invariant. Violation of universality, therefore, breaks neither gauge nor local Lorentz invariance.

Comment 11.2 Notice that, due to the gauge structure of Teleparallel Gravity, the action assumes a form similar to the action of a charged particle in an electromagnetic field. In fact, if the particle has additionally an electric charge q and is in the presence of an electromagnetic field A_μ, the action becomes

$$\mathscr{S} = -m_i c \int_{\mathsf{p}}^{\mathsf{q}} \left(\overset{\bullet}{\mathscr{D}}_\mu x^a u_a u^\mu + \frac{m_g}{m_i} B^a{}_\mu u_a u^\mu + \frac{q}{m_i} \frac{A_\mu u^\mu}{c^2} \right) ds. \tag{11.32}$$

We see from this expression that the gravitational mass m_g plays a role similar to the electric charge q. We see furthermore that, whereas the electromagnetic force is linear in the four-velocity, the gravitational force is quadratic.

Action (11.30) can be rewritten in the form

$$\mathscr{S} = -m_i c \int_{\mathsf{p}}^{\mathsf{q}} \big[u_a h^a{}_\mu + (\sqrt{\alpha_u} - 1) u_a B^a{}_\mu\big] \, dx^\mu, \tag{11.33}$$

where

$$h^a{}_\mu = \partial_\mu x^a + \overset{\bullet}{A}{}^a{}_{b\mu} x^b + B^a{}_\mu. \tag{11.34}$$

Comment 11.3 It is important to remark that, whereas $h^a{}_\mu$ above is a tetrad,

$$\tilde{h}^a{}_\mu = \tilde{h}_\mu x^a \tag{11.35}$$

is not because, by definition, a tetrad cannot depend on any property of the particle.

Up to the inertial mass outside the integration sign, the first term of the action (11.33) coincides with the universal action (6.33), whose equation of motion was found to be given by (6.50). The second term actually isolates the effects produced by the non-universality of gravity, as it vanishes when $m_i = m_g$. In order to obtain the non-universal equation of motion, we have to compute only the contribution from that second term.

To begin with we note that the variation of the four-velocity u_a can be written in the form

$$\delta u_a = \frac{du_a}{ds}\delta s. \tag{11.36}$$

On the other hand, from the relation $ds = g_{\mu\nu}u^\mu dx^\nu$, we can write

$$\delta s = g_{\mu\nu}u^\mu \delta x^\nu. \tag{11.37}$$

Substituting in (11.36), we obtain

$$\delta u_a = u_\mu \frac{du_a}{ds}\delta x^\mu. \tag{11.38}$$

Using this result, as well as the relations

$$\delta dx^\mu = d\delta x^\mu, \qquad \delta B^a{}_\mu = \partial_\rho B^a{}_\mu \delta x^\rho, \tag{11.39}$$

and taking into account that the variation of the first term of the action (11.33) yields the equation of motion (6.50), the variation of the whole action yields

$$\frac{du_a}{ds} - \overset{\bullet}{A}{}^b{}_{a\rho}u_b u^\rho = -\overset{\bullet}{K}{}^b{}_{a\rho}u_b u^\rho + F_a, \tag{11.40}$$

where

$$F_a = -(\sqrt{\alpha_u} - 1)h_a{}^\mu \left[P^\rho{}_\mu B^b{}_\rho \frac{du_b}{ds} - (\partial_\mu B^b{}_\rho - \partial_\rho B^b{}_\mu)u_b u^\rho\right] \tag{11.41}$$

is a new gravitational force, with

$$P^\rho{}_\mu = \delta^\rho_\mu - u^\rho u_\mu \tag{11.42}$$

a velocity-projection tensor. The first term on the right-hand side of (11.40) represents the universal gravitational force. The second term, on the other hand, represents the effects coming from the lack of universality. This means that the breaking of the weak equivalence principle would correspond to the discovery of a new force, not predicted by General Relativity. If $m_g = m_i$, this new force vanishes and the equation of motion reduces to

$$\frac{du_a}{ds} - \overset{\bullet}{A}{}^b{}_{a\rho}u_b u^\rho = -\overset{\bullet}{K}{}^b{}_{a\rho}u_b u^\rho, \tag{11.43}$$

which is the universal teleparallel force equation (6.50).

Observe that the universal and the non-universal parts of the gravitational force on the right-hand side of the equation of motion (11.40) are separately orthogonal to the four-velocity. Observe also that, although that equation of motion depends explicitly on the property

$$\sqrt{\alpha_u} = m_g/m_i$$

of the particle, the gauge potential $B^a{}_\mu$ does not. This means that the teleparallel field equation (9.58) can be consistently solved for $B^a{}_\mu$, independently of the validity or not of the weak equivalence principle—and the result inserted into (11.40)–(11.41). Stating the conclusion rather solemnly: Teleparallel Gravity is able to describe the motion of a particle even in the lack of universality [4].

Comment 11.4 Using the same procedure of Sect. 6.3, the newtonian limit of the equation of motion (11.40)–(11.41) is found to be

$$m_i \frac{d^2\boldsymbol{x}}{dt^2} = -m_g \nabla \Phi, \tag{11.44}$$

with

$$\Phi = c^2 B_{00}$$

the newtonian potential. This is just Newton's equation for $m_i \neq m_g$. We recall that newtonian gravity, like Teleparallel Gravity, can comply with the lack of universality. We see once more that the newtonian limit follows much more naturally from Teleparallel Gravity than from General Relativity.

11.3.3 Global Formulation and the COW Experiment

In Chap. 7 a global formulation for gravity was developed. In this section we are going to develop the same formulation, but in the lack of universality. In this case, the action integral describing the interaction of a particle of gravitational mass m_g with gravitation is [see Eq. (11.30)],

$$\mathcal{S}_g = \int_{\mathsf{p}}^{\mathsf{q}} m_i c u_a \sqrt{\alpha_u} B^a{}_\mu dx^\mu = \int_{\mathsf{p}}^{\mathsf{q}} m_g c B^a{}_\mu u_a \, dx^\mu. \tag{11.45}$$

The corresponding gravitational non-integrable phase factor is then

$$\Phi_g(\mathsf{p}|\mathsf{q}) = \exp\left(\frac{i m_g c}{\hbar} \int_{\mathsf{p}}^{\mathsf{q}} B^a{}_\mu u_a \, dx^\mu\right). \tag{11.46}$$

Similarly to the electromagnetic phase factor, it represents the *quantum* mechanical law that replaces the *classical* gravitational Lorentz force equation (11.40).

As an application of the gravitational non-integrable phase factor (11.46), we are going to reconsider the COW experiment studied in Sect. 7.2, but now in absence of universality. As already discussed, it consists in using a neutron interferometer to observe the quantum mechanical phase shift of neutrons caused by their interaction with Earth's gravitational field, which is usually assumed to be newtonian. Furthermore, as the experience is performed with thermal neutrons, it is possible to use the small velocity approximation. In this case, the gravitational phase factor (7.8) becomes

$$\Phi_g(\mathsf{p}|\mathsf{q}) = \exp\left(\frac{i m_g c^2}{\hbar} \int_{\mathsf{p}}^{\mathsf{q}} B_{00}\, dt\right), \tag{11.47}$$

where we have used the fact that $u^0 = \gamma \simeq 1$ for thermal neutrons. In the newtonian approximation, we can set

$$c^2 B_{00} \equiv \phi = g z, \tag{11.48}$$

with ϕ the (homogeneous) Earth newtonian potential. In this expression, g is the gravitational acceleration, assumed not to change significantly in the region of the experience, and z is the distance from Earth taken from some reference point. Consequently, the phase factor can be rewritten in the form

$$\Phi_g(\mathsf{p}|\mathsf{q}) = \exp\left(\frac{i m_g g}{\hbar} \int_{\mathsf{p}}^{\mathsf{q}} z(t)\, dt\right) \equiv \exp i\varphi. \tag{11.49}$$

Following the same steps used in Sect. 7.2, the gravitationally induced phase difference between the two trajectories of Fig. 7.1 is

$$\Delta\varphi \equiv \varphi_{\mathsf{bce}} - \varphi_{\mathsf{bde}} = \frac{m_g g r}{\hbar} \int_{\mathsf{c}}^{\mathsf{e}} dt. \tag{11.50}$$

Since the neutron velocity is constant along the segment ce, we have

$$\int_{\mathsf{c}}^{\mathsf{e}} dt \equiv \frac{s}{v} = \frac{s m_i \lambda}{h}, \tag{11.51}$$

where s is the length of the segment ce, and $\lambda = h/(m_i v)$ is the de Broglie wavelength of the neutron. In absence of universality, therefore, the gravitationally induced phase difference predicted for the COW experience is found to be [5]

$$\Delta\varphi = s\frac{2\pi g r \lambda m_i^2}{h^2}\sqrt{\alpha_u}. \tag{11.52}$$

If $m_g > m_i$, the phase difference would be larger than the universal effect; for $m_g < m_i$, the phase difference would be smaller than the universal effect. When the gravitational and inertial masses coincide, $m_g = m_i \equiv m$, the phase shift becomes

$$\Delta\varphi = s\frac{2\pi g r \lambda m^2}{h^2}, \tag{11.53}$$

which is the usual result obtained for the COW experiment [6, 7].

11.4 Non-universality and General Relativity

From Eqs. (11.22) and (11.34), we can write

$$\tilde{h}^c{}_\rho - h^c{}_\rho = (\sqrt{\alpha_u} - 1) B^c{}_\rho. \tag{11.54}$$

Using this relation, the equation of motion (11.40)–(11.41) can be rewritten, after some algebraic manipulations, in the form

$$\tilde{h}^a{}_\rho P^\lambda{}_\mu \left(\delta^\rho_\lambda \frac{\overset{\bullet}{\mathscr{D}} u_a}{\mathscr{D}s} - \overset{\bullet}{A}{}^b{}_{a\lambda} u_b u^\rho \right) = \sqrt{\alpha_u} \overset{\bullet}{T}{}^a{}_{\mu\rho} u_a u^\rho. \tag{11.55}$$

It is not a geodesic equation for any spacetime metric. In order to comply with the geometric foundations of General Relativity, and write it as a geodesic equation, it is necessary to incorporate the particle properties into the spacetime geometry. This can be achieved by assuming that $\tilde{h}^a{}_\mu$, which takes into account the property

$$\sqrt{\alpha_u} = m_g/m_i$$

of the particle, is a tetrad field [see Comment 11.3]. This "tetrad" defines a new spacetime metric tensor

$$\tilde{g}_{\mu\nu} = \eta_{ab} \tilde{h}^a{}_\mu \tilde{h}^b{}_\nu, \tag{11.56}$$

in terms of which the corresponding spacetime invariant interval is

$$d\tilde{s}^2 = \tilde{g}_{\mu\nu}\, dx^\mu\, dx^\nu. \tag{11.57}$$

The torsion related to the "tetrad" (11.22) is

$$\tilde{T}^a{}_{\mu\rho} = \partial_\mu \tilde{h}^a{}_\rho - \partial_\rho \tilde{h}^a{}_\mu + \overset{\bullet}{A}{}^a{}_{b\mu} \tilde{h}^b{}_\rho - \overset{\bullet}{A}{}^a{}_{b\rho} \tilde{h}^b{}_\mu. \tag{11.58}$$

Comparing with the definition of the teleparallel torsion,

$$\overset{\bullet}{T}{}^a{}_{\mu\rho} = \partial_\mu B^a{}_\rho - \partial_\rho B^a{}_\mu + \overset{\bullet}{A}{}^a{}_{b\mu} B^b{}_\rho - \overset{\bullet}{A}{}^a{}_{b\rho} B^b{}_\mu, \tag{11.59}$$

we see that

$$\sqrt{\alpha_u} \overset{\bullet}{T}{}^a{}_{\mu\rho} = \tilde{h}^a{}_\lambda \tilde{T}^\lambda{}_{\mu\rho}. \tag{11.60}$$

Using this result one can verify that, for a fixed relation m_g/m_i, the equation of motion (11.55) is equivalent to the geodesic equation

$$\frac{d\tilde{u}_\mu}{d\tilde{s}} - \tilde{\Gamma}^\lambda{}_{\mu\rho} \tilde{u}_\lambda \tilde{u}^\rho = 0, \tag{11.61}$$

where

$$\tilde{u}^\mu \equiv \frac{dx^\mu}{d\tilde{s}} = u^a \tilde{h}_a{}^\mu \tag{11.62}$$

is the particle four-velocity, and

$$\tilde{\Gamma}^\lambda{}_{\mu\rho} = \tfrac{1}{2} \tilde{g}^{\lambda\nu} (\partial_\mu \tilde{g}_{\nu\rho} + \partial_\rho \tilde{g}_{\mu\nu} - \partial_\nu \tilde{g}_{\mu\rho}) \tag{11.63}$$

is the Christoffel connection of the metric (11.56). Equation (11.61) can also be obtained from a variational principle, with the action integral given by

$$\tilde{\mathscr{S}} = -m_i c \int_{\mathrm{p}}^{\mathrm{q}} d\tilde{s}, \tag{11.64}$$

which has the form of the usual action in the context of General Relativity.

Nevertheless, the price for imposing a geodesic equation of motion to describe a non-universal interaction is that the theory becomes inconsistent. In fact, the solution of the corresponding Einstein's field equation

$$\tilde{R}_{\mu\nu} - \frac{1}{2}\tilde{g}_{\mu\nu}\tilde{R} = \frac{8\pi G}{c^4}\tilde{\Theta}_{\mu\nu}, \qquad (11.65)$$

would in this case depend on the relation m_g/m_i of the test particle, which renders the theory self-contradictory: test particles with different relations m_g/m_i would require connections with different curvatures to keep all equations of motion given by geodesics. More simply, different particles would "feel" different metrics (11.56). Of course the gravitational field, as a true field, cannot depend on the properties of any test particle. We can then conclude that, in the lack of the weak equivalence principle, the geometric description of General Relativity breaks down.

At the quantum level, deep conceptual changes occur with respect to classical gravity, the most important being the fact that gravitation seems to be no more universal [8, 9]. In fact, at this level, the phase of the particle wave function acquires a fundamental status, and turns out to depend on the particle mass. In principle, therefore, due to the lack of universality at the quantum level, the geometric description of General Relativity is unable to deal with gravitationally-induced quantum effects. Of course, since in the specific case of the COW experiment gravitation enters only in the newtonian form, the inability of General Relativity to deal with non-universal gravity does not show up. On the other hand, since the gauge approach of Teleparallel Gravity is able to deal with the gravitational interaction even in the lack of universality, this theory can be considered more appropriate to study gravitationally-induced quantum effects.

References

1. Will, C.M.: Living Rev. Relativ. **9**, 3 (2006). gr-qc/0510072
2. Weinberg, S.: The Quantum Theory of Fields, vol. II, p. 4, footnote. Cambridge University Press, Cambridge (1996)
3. Barrow, J.D., Tipler, F.J.: The Anthropic Cosmological Principle. Oxford University Press, Oxford (1988)
4. Aldrovandi, R., Pereira, J.G., Vu, K.H.: Gen. Relativ. Gravit. **36**, 101 (2004). gr-qc/0304106
5. Aldrovandi, R., Pereira, J.G., Vu, K.H.: Selected topics in teleparallel gravity. Braz. J. Phys. **34**, 1374 (2004). gr-qc/0312008 (Talk presented at the 24th National Meeting of the Brazilian Physical Society, Caxambú MG, Brazil, 2003)
6. Overhauser, A.W., Colella, R.: Phys. Rev. Lett. **33**, 1237 (1974)
7. Colella, R., Overhauser, A.W., Werner, S.A.: Phys. Rev. Lett. **34**, 1472 (1975)
8. Greenberger, D.: Ann. Phys. **47**, 116 (1968)
9. Greenberger, D., Overhauser, A.W.: Rev. Mod. Phys. **51**, 43 (1979)

Chapter 12
Gravitational Coupling of the Fundamental Fields

The teleparallel coupling of some of the main fundamental relativistic fields—scalar, Dirac and electromagnetic fields—to gravitation are examined.

12.1 Representations of the Lorentz Group

Each field (or particle) appearing in Nature can be assigned to a fixed representation of the Poincaré group [1, 2], the semi-direct product of the Lorentz and the translation groups. The Poincaré group is of rank two, which means that two independent invariant operators are defined which commute with every generator, and whose eigenvalues are characteristic of the representation—and in consequence of the field (or particle) assigned to it. As arbitrary functions of such invariants are also invariant, it turns out that a convenient choice is possible, such that one invariant is related to the translational sector and the other to the Lorentz sector. The eigenvalue coming from the translational part defines the mass of the field, whereas the eigenvalue coming from the Lorentz generators defines its spin. The commutation relations of the six Lorentz generators, which we denote here by $\boldsymbol{a}$ and $\boldsymbol{b}$, can be written in the form [3]

$$[a_i, a_j] = \varepsilon_{ijk} a_k \tag{12.1}$$

$$[b_i, b_j] = \varepsilon_{ijk} b_k \tag{12.2}$$

$$[a_i, b_j] = 0, \tag{12.3}$$

with $i, j, k, \ldots = 1, 2, 3$. A general spin-$\mathsf{s}$ representation of the Lorentz group can be constructed as either a field transforming under an irreducible representation, or as a direct sum of irreducible representations, each characterized by an integer or half-integer A and B, with

$$\boldsymbol{a}^2 = A(A+1) \quad \text{and} \quad \boldsymbol{b}^2 = B(B+1). \tag{12.4}$$

These representations are labeled by the numbers (A, B), where $\mathsf{s} = A + B$. The number of components n of the representation (A, B) is

$$n = (2A+1)(2B+1). \tag{12.5}$$

R. Aldrovandi, J.G. Pereira, *Teleparallel Gravity*, Fundamental Theories of Physics 173, DOI 10.1007/978-94-007-5143-9_12, © Springer Science+Business Media Dordrecht 2013

When $A \neq B$, the irreducible representation can be written as a direct sum of the form $(A, B) \oplus (B, A)$, with the number of independent components given by $2n$.

12.2 Gravitational Coupling Revisited

Let us repeat in other terms: a *relativistic field* is a field with a definite, covariant behavior under transformations of the Poincaré group $\mathscr{P}$ [more about that in Appendix D]. We say that it *belongs to* (or *live in*) some well-defined representation of $\mathscr{P}$. There is one of such representations for each value of mass and spin. With fixed masses, a scalar field (spin 0) belongs to the so-called "null" representation, which means that it is invariant. A Dirac field (spin $1/2$) belongs to the "bi-spinor" representation. A vector field (spin 1) belongs to the "vector" representation. And so on.

The teleparallel coupling of a general field Ψ to gravitation is obtained by applying the prescription [see Chap. 5]

$$\partial_\mu \Psi \to \ddot{\mathscr{D}}_\mu \Psi, \tag{12.6}$$

where

$$\ddot{\mathscr{D}}_\mu \Psi = \partial_\mu \Psi - \tfrac{i}{2}\big(\dot{A}^{bc}{}_\mu - \dot{K}^{bc}{}_\mu\big) S_{bc} \Psi \tag{12.7}$$

is the teleparallel Fock-Ivanenko covariant derivative, with S_{bc} the Lorentz generators written in the representation to which the field Ψ is assigned. Concomitantly, the tetrad and the metric must also be replaced by their gravitational counterparts,

$$e^a{}_\mu \to h^a{}_\mu \quad \text{and} \quad \eta_{\mu\nu} \to g_{\mu\nu}. \tag{12.8}$$

Of course, due to the identity

$$\dot{A}^{bc}{}_\mu - \dot{K}^{bc}{}_\mu = \mathring{A}^{bc}{}_\mu, \tag{12.9}$$

the recipe above is equivalent to the General Relativity coupling prescription

$$\partial_\mu \Psi \to \mathring{\mathscr{D}}_\mu \Psi, \tag{12.10}$$

with

$$\mathring{\mathscr{D}}_\mu \Psi = \partial_\mu \Psi - \tfrac{i}{2} \mathring{A}^{bc}{}_\mu S_{bc} \Psi \tag{12.11}$$

the usual Fock-Ivanenko covariant derivative. Although equivalent, however, it is conceptually different, in the sense that the interpretation of the gravitational interaction is different in each case, as discussed in Chap. 5. In the present chapter we apply the teleparallel coupling prescription to some specific fundamental fields.

12.3 Scalar Field

The scalar field ϕ is associated with the one-component representation $(0,0)$. In Minkowski spacetime, its lagrangian is written as

$$\mathscr{L}_\phi = \frac{e}{2}\big(\eta^{\mu\nu}\partial_\mu\phi\partial_\nu\phi - \mu^2\phi^2\big), \tag{12.12}$$

where $e = \det(e^a{}_\mu) = 1$ and $\mu = mc/\hbar$, with m the mass of the field. The corresponding field equation is the Klein-Gordon equation

$$\Box\phi + \mu^2\phi = 0, \tag{12.13}$$

with $\Box = \eta^{\mu\nu}\partial_\mu\partial_\nu$ the d'Alembertian operator. Since by definition a scalar field belongs to the null representation,

$$S_{ab}\phi = 0, \tag{12.14}$$

the coupling prescription (12.6) assumes the form

$$\partial_\mu \to \ddot{\mathscr{D}}_\mu \equiv \partial_\mu. \tag{12.15}$$

Applying this prescription to the free lagrangian (12.12), it becomes

$$\mathscr{L}_\phi = \frac{h}{2}\big(g^{\mu\nu}\partial_\mu\phi\partial_\nu\phi - \mu^2\phi^2\big), \tag{12.16}$$

where we have concomitantly changed $e^a{}_\mu \to h^a{}_\mu$ and $\eta^{\mu\nu} \to g^{\mu\nu}$. Variation of this lagrangian yields the field equation

$$\ddot{\Box}\phi + \mu^2\phi = 0, \tag{12.17}$$

where

$$\ddot{\Box}\phi = h^{-1}\partial_\rho\big(h\partial^\rho\phi\big) \tag{12.18}$$

is the Laplace-Beltrami operator. Using the identity

$$\partial_\mu h = h\big(\dot{\Gamma}^\rho{}_{\mu\rho} - \dot{K}^\rho{}_{\mu\rho}\big), \tag{12.19}$$

it can be rewritten in the form

$$\ddot{\Box}\phi = \big(\partial_\mu + \dot{\Gamma}^\rho{}_{\mu\rho} - \dot{K}^\rho{}_{\mu\rho}\big)\partial^\mu\phi \equiv \ddot{\nabla}_\mu\partial^\mu\phi. \tag{12.20}$$

This is the teleparallel Laplace-Beltrami operator. The corresponding Klein-Gordon equation is, consequently [4],

$$\ddot{\nabla}_\mu\partial^\mu\phi + \mu^2\phi = 0. \tag{12.21}$$

We see from Eq. (12.20) that, although the scalar field belongs to the null representation of the Lorentz group, it couples to torsion—or equivalently, to contortion—through its derivative.

Comment 12.1 We remark that, in the Einstein-Cartan models [see Chap. 17], only a spin distribution could produce or feel torsion [5]. A scalar field, for example, should be able to feel only curvature [6]. However, from the point of view of the teleparallel equivalent of General Relativity, since a scalar field couples to curvature, it must also couple to torsion [4]. Field equation (12.21) shows that this is in fact the case.

12.4 Dirac Spinor Field

The spinor representations of the Lorentz group are the most fundamental ones. The $(1/2, 0)$ representation has spin $1/2$ and describes a (let us say) left-handed Weyl spinor. The $(0, 1/2)$ representation describes a right-handed Weyl spinor. The linear combination

$$(1/2, 0) \oplus (0, 1/2) \tag{12.22}$$

describes a Dirac (bi-)spinor. The reason why the spinor representations are the most fundamental is that they can be used to construct, by multiplying them together, any other representation of the Lorentz group. For example, the scalar representation $(0, 0)$ can be obtained from the antisymmetric product of two $(1/2, 0)$ representations:

$$\left[(1/2, 0) \otimes (1/2, 0)\right]_a = (0, 0). \tag{12.23}$$

12.4.1 *The Dirac Equation*

On Minkowski spacetime, the spinor field lagrangian is

$$\mathscr{L}_\psi = e\left[\frac{ic\hbar}{2}\left(\bar{\psi}\gamma^a e_a{}^\mu \partial_\mu \psi - e_a{}^\mu \partial_\mu \bar{\psi}\gamma^a \psi\right) - mc^2 \bar{\psi}\psi\right], \tag{12.24}$$

with $e = \det(e^a{}_\mu) = 1$. The corresponding field equation is the free Dirac equation

$$i\hbar\gamma^a e_a{}^\mu \partial_\mu \psi - mc\psi = 0. \tag{12.25}$$

Comment 12.2 Due to the importance of spinor fields in the study of the gravitational interaction at the microscopic scale, in Appendix D we present a résumé on the Dirac equation.

The gravitationally coupled Dirac lagrangian is obtained by applying the teleparallel coupling prescription

$$e_a{}^\mu \partial_\mu \psi \to h_a{}^\mu \ddot{\mathscr{D}}_\mu = \partial_\mu \psi - \tfrac{i}{2}\left(\dot{A}^{bc}{}_\mu - \dot{K}^{bc}{}_\mu\right) S_{bc}\psi, \tag{12.26}$$

with S_{bc} the Lorentz generators in the spinor representation:

$$S_{bc} \equiv \tfrac{1}{2}\sigma_{bc} = \tfrac{i}{4}[\gamma_b, \gamma_c]. \tag{12.27}$$

The result is

$$\mathscr{L}_\psi = h\left[\frac{ic\hbar}{2}\left(\bar{\psi}\gamma^\mu \ddot{\mathscr{D}}_\mu \psi - \ddot{\mathscr{D}}_\mu \bar{\psi}\gamma^\mu \psi\right) - mc^2 \bar{\psi}\psi\right], \tag{12.28}$$

where we have denoted $\gamma^\mu \equiv \gamma^\mu(x) = \gamma^a h_a{}^\mu$. Using the identity

$$\ddot{\mathscr{D}}_\mu\left(h\gamma^\mu\right) = 0, \tag{12.29}$$

the teleparallel version of the coupled Dirac equation is found to be

$$i\hbar\gamma^\mu\left[\partial_\mu \psi - \tfrac{i}{4}\left(\dot{A}^{bc}{}_\mu - \dot{K}^{bc}{}_\mu\right)\sigma_{bc}\right]\psi - mc\psi = 0. \tag{12.30}$$

In the class of frames in which the inertial connection $\overset{\bullet}{A}{}^{ab}{}_{\mu}$ vanishes, it becomes

$$i\hbar\gamma^{\mu}\left[\partial_{\mu}\psi+\tfrac{i}{4}\overset{\bullet}{K}{}^{bc}{}_{\mu}\sigma_{bc}\right]\psi-mc\psi=0. \tag{12.31}$$

12.4.2 *Torsion Decomposition and Spinors*

As discussed in Sect. 1.7, torsion can be decomposed into irreducible components under the global Lorentz group:

$$T_{\lambda\mu\nu}=\tfrac{2}{3}\left(\overset{\bullet}{\mathscr{T}}_{\lambda\mu\nu}-\overset{\bullet}{\mathscr{T}}_{\lambda\nu\mu}\right)+\tfrac{1}{3}\left(g_{\lambda\mu}\overset{\bullet}{\mathscr{V}}_{\nu}-g_{\lambda\nu}\overset{\bullet}{\mathscr{V}}_{\mu}\right)+\varepsilon_{\lambda\mu\nu\rho}\overset{\bullet}{\mathscr{A}}{}^{\rho}, \tag{12.32}$$

where $\overset{\bullet}{\mathscr{T}}_{\lambda\mu\nu}$ represents the purely tensor part, and $\overset{\bullet}{\mathscr{V}}_{\mu}$ and $\overset{\bullet}{\mathscr{A}}{}^{\rho}$ represent respectively its vector and axial parts. Let us then consider the Dirac equation (12.31). A simple calculation shows that the coupling term of the covariant derivative is

$$\tfrac{i}{4}\overset{\bullet}{K}{}^{ab}{}_{\mu}\gamma^{\mu}\sigma_{ab}=-\gamma^{\mu}\left(\tfrac{1}{2}\overset{\bullet}{\mathscr{V}}_{\mu}+\tfrac{3i}{4}\overset{\bullet}{\mathscr{A}}_{\mu}\gamma^{5}\right), \tag{12.33}$$

where

$$\gamma^{5}=\gamma_{5}:=i\gamma^{0}\gamma^{1}\gamma^{2}\gamma^{3}. \tag{12.34}$$

As a consequence, in the class of frames in which the inertial connection $\overset{\bullet}{A}{}^{ab}{}_{\mu}$ vanishes, the teleparallel covariant derivative of a spinor field can be written as

$$\gamma^{\mu}\overset{\bullet\bullet}{\mathscr{D}}_{\mu}\psi=\gamma^{\mu}\left(\partial_{\mu}-\tfrac{1}{2}\overset{\bullet}{\mathscr{V}}_{\mu}-\tfrac{3i}{4}\overset{\bullet}{\mathscr{A}}_{\mu}\gamma^{5}\right)\psi. \tag{12.35}$$

It involves the vector $\overset{\bullet}{\mathscr{V}}_{\mu}$ and the axial $\overset{\bullet}{\mathscr{A}}_{\mu}$ torsions only [7]. This means essentially that the purely tensor piece $\overset{\bullet}{\mathscr{T}}_{\lambda\mu\nu}$ of torsion is irrelevant for the description of the gravitational interaction of fermions. Notice that this property is necessary for the invariance of the lagrangian (12.28) under time reversal T and space inversion P. In fact, whereas vector $\overset{\bullet}{\mathscr{V}}_{\mu}$ couples to the vector current γ^{μ}, axial $\overset{\bullet}{\mathscr{A}}_{\mu}$ couples to the axial current $\gamma^{\mu}\gamma^{5}$. Of course, both couplings are invariant under a combined operation PT. Since they are also invariant under charge conjugation C, they are CPT invariant. The corresponding Dirac equation

$$i\hbar\gamma^{\mu}\left(\partial_{\mu}-\tfrac{1}{2}\overset{\bullet}{\mathscr{V}}_{\mu}-\tfrac{3i}{4}\overset{\bullet}{\mathscr{A}}_{\mu}\gamma^{5}\right)\psi=mc\psi \tag{12.36}$$

is consequently also invariant.

Comment 12.3 It is interesting to remark that, in General Relativity, where the covariant derivative of a spinor field is written as

$$\overset{\circ}{\mathscr{D}}_{\mu}\psi=\partial_{\mu}\psi-\tfrac{i}{2}\overset{\circ}{A}{}^{ab}{}_{\mu}S_{ab}\psi, \tag{12.37}$$

if the spin connection $\overset{\circ}{A}{}^{bc}{}_{\mu}$ is expressed in terms of the coefficient of nonholonomy $f^{a}{}_{bc}$ through

$$\overset{\circ}{A}{}^{a}{}_{bc}=-\tfrac{1}{2}\left(f^{a}{}_{bc}+f_{bc}{}^{a}+f_{cb}{}^{a}\right), \tag{12.38}$$

a decomposition similar to (12.33) can be made, and the Dirac equation turns out to be written in terms of the trace and the pseudo-trace of $f^{a}{}_{bc}$ only. The purely tensor part of $f^{a}{}_{bc}$ is also irrelevant for spinors.

12.5 Electromagnetic Field

The electromagnetic vector potential A_μ transforms according to the representation

$$(1/2, 1/2) \tag{12.39}$$

of the Lorentz group, which describes a spin-1 field with $n = 4$ components. The electromagnetic field strength

$$F_{\mu\nu} = \partial_\mu A_\nu - \partial_\nu A_\mu, \tag{12.40}$$

with $n = 6$ components, transforms under the spin-1 representation

$$(1, 0) \oplus (0, 1) \tag{12.41}$$

of the Lorentz group. The representation $(1, 0)$ can be identified with an antisymmetric, self-dual second-rank tensor. Analogously, the representation $(0, 1)$ can be identified with an antisymmetric, anti-self-dual second-rank tensor. The representation (12.41), therefore, describes a parity invariant 2-form field. By construction, the field strength $F_{\mu\nu}$ satisfies the geometric Bianchi identity

$$\partial_\mu F_{\nu\rho} + \partial_\rho F_{\mu\nu} + \partial_\nu F_{\rho\mu} = 0 \tag{12.42}$$

which constitutes the first pair of Maxwell's equations.

On Minkowski spacetime, the electromagnetic field is described by the lagrangian density

$$\mathscr{L}_{em} = -\frac{e}{4} F_{\mu\nu} F^{\mu\nu}, \tag{12.43}$$

where $e = \det(e^a{}_\mu) = 1$. The corresponding field equation is

$$\partial_\mu F^{\mu\nu} = 0, \tag{12.44}$$

which constitutes the second pair of Maxwell's equations. Together with the Bianchi identity (12.42), they constitute the whole set of sourceless Maxwell's equations. In the Lorenz gauge $\partial_\mu A^\mu = 0$, field equation (12.44) becomes the wave equation

$$\Box A^\mu = 0, \tag{12.45}$$

with $\Box = \eta^{\mu\nu} \partial_\mu \partial_\nu$ the d'Alembertian operator.

Let us obtain now the gravitationally coupled Maxwell's equation in the context of Teleparallel Gravity. To begin with we note that, in the specific case of the electromagnetic vector field A^ρ, the coupling prescription (12.6) assumes the form

$$\partial_\mu A^\rho \to \ddot{\nabla}_\mu A^\rho = \partial_\mu A^\rho + \left(\dot{\Gamma}^\rho{}_{\nu\mu} - \dot{K}^\rho{}_{\nu\mu}\right) A^\nu. \tag{12.46}$$

Using the explicit form of $\ddot{\nabla}_\mu$, and the definitions of torsion and contortion tensors, it is easy to verify that the field strength

$$F_{\mu\nu} = \ddot{\nabla}_\mu A_\nu - \ddot{\nabla}_\nu A_\mu, \tag{12.47}$$

like in General Relativity, does not change:

$$F_{\mu\nu} = \partial_\mu A_\nu - \partial_\nu A_\mu. \tag{12.48}$$

In consequence, the Bianchi identity (12.42) remains unchanged:

$$\partial_\mu F_{\nu\sigma} + \partial_\sigma F_{\mu\nu} + \partial_\nu F_{\sigma\mu} = 0. \tag{12.49}$$

By replacing $e \to h$, with $h = \det(h^a{}_\mu)$, the gravitationally-coupled Maxwell lagrangian in Teleparallel Gravity is written as

$$\mathscr{L}_{em} = -\frac{h}{4} F_{\mu\nu} F^{\mu\nu}. \tag{12.50}$$

The corresponding field equation is

$$\ddot{\nabla}_\mu F^{\mu\nu} \equiv \partial_\mu \left(h F^{\mu\nu} \right) = 0, \tag{12.51}$$

which is the second pair of Maxwell's equation in Teleparallel Gravity. In terms of the electromagnetic potential, it reads

$$\ddot{\nabla}_\mu \left(\ddot{\nabla}^\mu A^\nu - \ddot{\nabla}^\nu A^\mu \right) = 0. \tag{12.52}$$

Using the commutation relation

$$\left[\ddot{\nabla}_\mu, \ddot{\nabla}_\nu \right] A^\mu = -\dot{Q}_{\mu\nu} A^\mu, \tag{12.53}$$

where $\dot{Q}_{\mu\nu} = \dot{Q}^\rho{}_{\mu\rho\nu}$, with $\dot{Q}^\rho{}_{\mu\sigma\nu}$ the tensor given by Eq. (9.24), as well as the teleparallel Lorentz gauge

$$\ddot{\nabla}_\mu A^\mu = 0, \tag{12.54}$$

it reduces to

$$\ddot{\nabla}_\mu \ddot{\nabla}^\mu A_\nu + \dot{Q}^\mu{}_\nu A_\mu = 0. \tag{12.55}$$

This is the teleparallel version of the second pair of Maxwell's equation in terms of the electromagnetic potential.

In the context of Teleparallel Gravity, therefore, the electromagnetic field is able to couple to torsion, and this coupling does not violate the gauge invariance of Maxwell's theory. Furthermore, using relation (9.21), it is easy to verify that the teleparallel version of Maxwell's equations, which are equations written in terms of the Weitzenböck connection only, are completely equivalent to the usual Maxwell's equations in the context of General Relativity,

$$\mathring{\nabla}_\mu \mathring{\nabla}^\mu A_\nu - \mathring{R}^\mu{}_\nu A_\mu = 0, \tag{12.56}$$

which are equations written in terms of the Levi-Civita connection only [8].

References

1. Wigner, E.: Ann. Math. **40**, 39 (1939)
2. Bargmann, V., Wigner, E.P.: Proc. Natl. Acad. Sci. **34**, 211 (1946)
3. Weinberg, S.: The Quantum Theory of Fields, vol. 1, p. 230. Cambridge University Press, Cambridge (1995)
4. de Andrade, V.C., Pereira, J.G.: Gen. Relativ. Gravit. **30**, 263 (1998). gr-qc/9706070

5. de Sabbata, V., Sivaram, C.: Spin and Torsion in Gravitation. World Scientific, Singapore (1994)
6. Hehl, F.H., von der Heyde, P., Kerlick, G.D., Nester, J.M.: Rev. Mod. Phys. **48**, 393 (1976)
7. Mosna, R.A., Pereira, J.G.: Gen. Relativ. Gravit. **36**, 2525 (2004). gr-qc/0312093
8. de Andrade, V.C., Pereira, J.G.: Int. J. Mod. Phys. D **8**, 141 (1999). gr-qc/9708051

Chapter 13
Spin-2 Field Coupled to Gravitation

The old problem of the inconsistencies that appear when a fundamental spin-2 field is coupled to gravitation is reconsidered from the teleparallel point of view. Provided the very notion of a spin-2 find is changed, this problem can find a solution in the teleparallel context.

13.1 Conformal Transformations

Under a conformal re-scaling of the metric tensor,

$$\bar{g}_{\mu\nu} = \Omega^2 g_{\mu\nu}, \tag{13.1}$$

with $\Omega = \Omega(x)$ the conformal factor, a general field Ψ transforms according to

$$\bar{\Psi} = \Omega^{d\tau} \Psi, \tag{13.2}$$

where d is a matrix that multiplies every Fermi field by $3/2$ and every Bose field by 1, and τ is a real number that depends on the spin s of the field [1]. For fermions, it is given by $\tau = 2(\mathsf{s} - 1)$, in which case we obtain

$$\bar{\Psi} = \Omega^{3(\mathsf{s}-1)} \Psi. \tag{13.3}$$

For bosons, on the other hand, it is given by $\tau = \mathsf{s} - 1$, which yields

$$\bar{\Psi} = \Omega^{\mathsf{s}-1} \Psi. \tag{13.4}$$

The power of Ω is known as the *conformal weight* w of the field Ψ. A scalar field has $w = -1$, a vector field has $w = 0$, and a spin-2 field has $w = 1$.

13.2 Fundamental Spin-2 Field

The dynamics of a fundamental spin-2 field in Minkowski spacetime is expected to coincide with the dynamics of a linear perturbation $\zeta_{\mu\nu}$ of the metric around flat spacetime [2]:

$$g_{\mu\nu} = \eta_{\mu\nu} + \zeta_{\mu\nu}. \tag{13.5}$$

R. Aldrovandi, J.G. Pereira, *Teleparallel Gravity*, Fundamental Theories of Physics 173,
DOI 10.1007/978-94-007-5143-9_13,

For this reason, a fundamental spin-2 field is usually assumed to be described by a symmetric, second-rank tensor $\zeta_{\mu\nu} = \zeta_{\nu\mu}$. However, there are some problems with this assumption. If a spin-2 field were defined as a perturbation of the metric, it would have conformal weight $w = 2$, instead of the correct conformal weight $w = 1$. A possible solution to this problem is to introduce a scalar field ϕ, and assume that the metric perturbation $\zeta_{\mu\nu}$ and a fundamental spin-2 field $\psi_{\mu\nu}$ are related by

$$\psi_{\mu\nu} = \phi \zeta_{\mu\nu}. \tag{13.6}$$

In this case, under the conformal re-scaling of the metric (13.1), it transforms according to

$$\bar{\psi}_{\mu\nu} = \Omega \psi_{\mu\nu}, \tag{13.7}$$

as appropriate for a spin-2 field. This is actually a matter of necessity because, in addition to presenting the correct conformal weight, $\psi_{\mu\nu}$ has now the appropriate field dimension—remember that the metric, as well as its perturbation, are dimensionless. In terms of $\psi_{\mu\nu}$, Eq. (13.5) assumes the form

$$g_{\mu\nu} = \eta_{\mu\nu} + \sqrt{k} \psi_{\mu\nu}, \tag{13.8}$$

where, we recall, $k = 8\pi G/c^4$. This constant is necessary to match the dimensions between the metric and the spin-2 field.

Now, due to the fact that it describes the gravitational interaction through a geometrization of spacetime, General Relativity is not, strictly speaking, a field theory in the usual sense of classical fields. On the other hand, owing to its gauge structure, Teleparallel Gravity does not geometrize the gravitational interaction, and for this reason it is much more akin to a field theory than General Relativity. When looking for a field theory for the spin-2 field, therefore, it seems far more reasonable to use Teleparallel Gravity as a paradigm. Furthermore, conceptually speaking, a symmetric second-rank tensor is not the most fundamental notion of a spin-2 field. As is well known, although the gravitational interaction of tensor fields can be described in the metric formalism, the gravitational interaction of spinors requires a tetrad formalism [3]. The latter can then be considered to be more fundamental than the metric formulation, in the sense that it is able to describe the gravitational interaction of both tensor and spinor fields. And, accordingly, the tetrad field can be said to be more fundamental than the metric.

Relying on these arguments, instead of similar to a linear perturbation of the metric, a fundamental spin-2 field should be considered as similar to a linear perturbation of the tetrad field. Denoting by $e^a{}_\mu$ a trivial tetrad representing the Minkowski metric, such perturbation reads

$$h^a{}_\nu = e^a{}_\nu + \zeta^a{}_\nu, \tag{13.9}$$

where $\zeta^a{}_\nu$ are the components of a 1-form assuming values in the Lie algebra of the translation group,

$$\zeta_\nu = \zeta^a{}_\nu P_a, \tag{13.10}$$

with $P_a = \partial_a$ the translation generators. Like in the case of the metric perturbation, the field variable representing a fundamental translational-valued 1-form is

$$\phi^a{}_\nu = \phi \zeta^a{}_\nu, \tag{13.11}$$

with ϕ a scalar field. In this case, the tetrad perturbation assumes the form

$$h^a{}_\mu = e^a{}_\mu + \sqrt{k}\phi^a{}_\mu. \tag{13.12}$$

According to the teleparallel paradigm, therefore, instead of a symmetric second-rank tensor $\psi_{\mu\nu}$, a spin-2 field is assumed to be represented by a spacetime (world) vector field assuming values in the Lie algebra of the translation group

$$\phi_\mu = \phi^a{}_\mu P_a. \tag{13.13}$$

Its components $\phi^a{}_\mu$, like the gauge potential $B^a{}_\mu$ of Teleparallel Gravity, represent a set of four spacetime vector fields. Of course, as a 1-form, it is invariant under a conformal re-scaling of the metric:

$$\bar{\phi}^a{}_\nu = \phi^a{}_\nu. \tag{13.14}$$

Observe that, in the usual *metric* formulation of gravity, the symmetry of the metric tensor eliminates six degrees of freedom of the sixteen original degrees of freedom of $g_{\mu\nu}$. In the *tetrad* formulation, on the other hand, local Lorentz invariance is responsible for eliminating six degrees of freedom of the sixteen original degrees of freedom of $h^a{}_\mu$, yielding the same number of independent components of $g_{\mu\nu}$. Of course, the same equivalence must hold in relation to the fields $\psi_{\mu\nu}$ and $\phi^a{}_\nu$. Furthermore, in the case of massless fields, the corresponding field equations must present additional symmetries such that the ten degrees of freedom be reduced to only two—as appropriate for a massless field.

13.3 The Flat Spacetime Case

13.3.1 Gauge Transformations

In the inertial frame e'^a, the tetrad describing the flat Minkowski spacetime is of the form

$$e'^a{}_\mu = \partial_\mu x'^a. \tag{13.15}$$

A spin-2 field $\phi'^a{}_\mu$ corresponds to a linear perturbation of this tetrad,

$$h'^a{}_\mu = \partial_\mu x'^a + \sqrt{k}\phi'^a{}_\mu. \tag{13.16}$$

In this class of frames, therefore, the vacuum is represented by

$$\phi'^a{}_\mu = (1/\sqrt{k})\partial_\mu \xi^a(x), \tag{13.17}$$

with $\xi^a(x)$ an arbitrary function of the spacetime coordinates x^ρ. In fact, such $\phi'^a{}_\mu$ represents simply a gauge translation

$$x'^a \to x'^a + \xi^a(x) \tag{13.18}$$

in the fiber, or tangent space. This means that the gauge transformation associated to the spin-2 field $\phi^a{}_\mu$ is

$$\phi'^a{}_\mu \to \phi'^a{}_\mu - (1/\sqrt{k})\partial_\mu \xi^a(x). \tag{13.19}$$

Of course, $h'^a{}_\mu$ is invariant under such transformations; hence the name "gauge" transformations.

13.3.2 Field Strength and Bianchi Identity

The next step towards the construction of a field theory for $\phi'^a{}_\mu$ is to define the analogous of torsion:

$$F'^a{}_{\mu\nu} = \partial_\mu \phi'^a{}_\nu - \partial_\nu \phi'^a{}_\mu. \tag{13.20}$$

This tensor is actually the spin-2 field-strength. As can be easily verified, $F'^a{}_{\mu\nu}$ is gauge invariant. Furthermore, it satisfies the Bianchi identity

$$\partial_\rho F'^a{}_{\mu\nu} + \partial_\nu F'^a{}_{\rho\mu} + \partial_\mu F'^a{}_{\nu\rho} = 0, \tag{13.21}$$

which can equivalently be written in the form

$$\partial_\rho \left(\varepsilon^{\lambda\rho\mu\nu} F'^a{}_{\mu\nu}\right) = 0, \tag{13.22}$$

with $\varepsilon^{\lambda\rho\mu\nu}$ the totally anti-symmetric, flat spacetime Levi-Civita tensor.

13.3.3 Lagrangian and Field Equation

Considering that the dynamics of a spin-2 field must coincide with the dynamics of linear gravity, its lagrangian will be similar to the lagrangian (9.13) of teleparallel gravity. One has just to replace the teleparallel torsion by the spin-2 field strength,

$$\overset{\bullet}{T}{}^a{}_{\mu\nu} \to \sqrt{k} F'^a{}_{\mu\nu}. \tag{13.23}$$

When one does that, the spin-2 analogous of the teleparallel superpotential $\overset{\bullet}{S}{}_a{}^{\mu\nu}$ is the field excitation 2-form

$$\mathcal{F}'^{\mu\nu}_a = e'^{\ \rho}_a \mathcal{K}'^{\mu\nu}{}_\rho - e'^{\ \nu}_a e'^{\ \rho}_b F'^{b\mu}{}_\rho + e'^{\ \mu}_a e'^{\ \rho}_b F'^{b\nu}{}_\rho, \tag{13.24}$$

with

$$\mathcal{K}'^{\mu\nu}{}_\rho = \tfrac{1}{2}\left(e'^{\ \nu}_a F'^{a\mu}{}_\rho + e'^a{}_\rho F'^{\mu\nu}_a - e'^{\ \mu}_a F'^{a\nu}{}_\rho\right) \tag{13.25}$$

the spin-2 analogous of the teleparallel contortion. As $\det(e'^a{}_\mu) = 1$ in the absence of gravitation, the lagrangian for a massless spin-2 field is found to be

$$\mathcal{L}' = \frac{1}{4} F'^a{}_{\mu\nu} \mathcal{F}'^{\mu\nu}_a. \tag{13.26}$$

By performing variations with respect to $\phi'^a{}_\rho$, we obtain

$$\partial_\mu \mathscr{F}'^{\rho\mu}_a = 0. \tag{13.27}$$

This is the field equation satisfied by a massless spin-2 field in Minkowski spacetime, as seen from the inertial frame $e'^a{}_\mu$.

Comment 13.1 The field excitation (13.24) coincides formally with the Fierz tensor, which justifies our notation. The field equation (13.27), on the other hand, is the linearized gravitational field equation in the Fierz formalism [4, 5]. It is a remarkable property of teleparallel gravity that its linear version naturally yields the Fierz formulation for a spin-2 fundamental field [6].

13.3.4 Duality Symmetry

The spin-2 field can be viewed as an abelian gauge field with the *internal* index replaced by an *external* Lorentz index. As discussed in Chap. 8, the presence of a tetrad allows Lorentz and spacetime indices to be transformed into each other. As a consequence, its Hodge dual will necessarily include additional index contractions in relation to the usual dual. Taking into account all possible contractions, its dual turns out to be given by [7]

$$\star F'^a{}_{\mu\nu} = \tfrac{1}{2} \varepsilon_{\mu\nu\rho\sigma} \mathscr{F}'^{a\rho\sigma}. \tag{13.28}$$

Substituting the Fierz tensor (13.24), we find

$$\star F'^a{}_{\mu\nu} = \tfrac{1}{2} \varepsilon_{\mu\nu\rho\sigma} \left(F'^{a\rho\sigma} - e'^{a\sigma} e'_b{}^\lambda F'^{b\rho}{}_\lambda + e'^{a\rho} e'_b{}^\lambda F'^{b\sigma}{}_\lambda \right). \tag{13.29}$$

The Bianchi identity (13.22), written for the dual of $F'^a{}_{\mu\nu}$, reads

$$\partial_\rho \left(\varepsilon^{\lambda\rho\mu\nu} \star F'^a{}_{\mu\nu} \right) = 0. \tag{13.30}$$

Substituting $\star F'^a{}_{\mu\nu}$ as given by Eq. (13.28), we get

$$\partial_\rho \mathscr{F}'^{a\mu\rho} = 0, \tag{13.31}$$

which is the field equation (13.27). We see in this way that, provided the generalized Hodge dual (13.28) for soldered bundles is used, the spin-2 field exhibits the duality symmetry.

13.3.5 Passage to a General Frame

In a Lorentz rotated frame $e^a = \Lambda^a{}_b(x) e'^a$, the tetrad assumes the form

$$e^a{}_\mu \equiv \overset{\bullet}{\mathscr{D}}_\mu x^a = \partial_\mu x^a + \overset{\bullet}{A}{}^a{}_{b\mu} x^b. \tag{13.32}$$

In such a frame, the vacuum of $\phi^a{}_\mu$ turns out to be represented by

$$\phi^a{}_\mu = (1/\sqrt{k}) \overset{\bullet}{\mathscr{D}}_\mu \xi^a(x), \tag{13.33}$$

whereas the gauge transformations assume the form

$$\phi^a{}_\mu \to \phi^a{}_\mu - (1/\sqrt{k})\overset{\bullet}{\mathscr{D}}_\mu \xi^a(x). \tag{13.34}$$

The field strength (13.20), on the other hand, becomes

$$F^a{}_{\mu\nu} = \overset{\bullet}{\mathscr{D}}_\mu \phi^a{}_\nu - \overset{\bullet}{\mathscr{D}}_\nu \phi^a{}_\mu. \tag{13.35}$$

Accordingly, the Bianchi identity reads

$$\overset{\bullet}{\mathscr{D}}_\rho F^a{}_{\mu\nu} + \overset{\bullet}{\mathscr{D}}_\nu F^a{}_{\rho\mu} + \overset{\bullet}{\mathscr{D}}_\mu F^a{}_{\nu\rho} = 0, \tag{13.36}$$

which is equivalent to

$$\overset{\bullet}{\mathscr{D}}_\rho \left(\varepsilon^{\lambda\rho\mu\nu} F^a{}_{\mu\nu}\right) = 0. \tag{13.37}$$

The lagrangian (13.26) is, of course, invariant under local Lorentz transformations,

$$\mathscr{L}' \equiv \mathscr{L} = \frac{1}{4} \mathscr{F}_a{}^{\mu\nu} F^a{}_{\mu\nu}, \tag{13.38}$$

with the field excitation 2-form assuming now the form

$$\mathscr{F}_a{}^{\mu\nu} = e_a{}^\rho \mathscr{K}^{\mu\nu}{}_\rho - e_a{}^\nu e_b{}^\rho F^{b\mu}{}_\rho + e_a{}^\mu e_b{}^\rho F^{b\nu}{}_\rho. \tag{13.39}$$

The corresponding field equation, given by

$$\overset{\bullet}{\mathscr{D}}_\mu \mathscr{F}_a{}^{\rho\mu} \equiv \partial_\mu \mathscr{F}_a{}^{\rho\mu} - \overset{\bullet}{A}{}^b{}_{a\mu} \mathscr{F}_b{}^{\rho\mu} = 0, \tag{13.40}$$

represents the field equation satisfied by a massless spin-2 field in Minkowski spacetime, as seen from the general frame $e^a{}_\mu$. Observe that the theory has twenty two constraints: sixteen from the invariance under the gauge transformations (13.34), and six from the invariance of the lagrangian $\mathscr{L}$ under local Lorentz transformations. The twenty four original components of the field excitation $\mathscr{F}_a{}^{\mu\nu}$ are then reduced to only two, as suitable for a massless spin-2 field.

13.3.6 Relation to the Metric Approach

The relation between the spacetime and the tangent space metrics is given by

$$g_{\mu\nu} = \eta_{ab} h'^a{}_\mu h'^b{}_\nu. \tag{13.41}$$

Substituting $g_{\mu\nu}$ as given by (13.8), and $h'^a{}_\mu$ as given by (13.16), up to first-order in $\phi'^a{}_\mu$ we obtain

$$\psi^\mu{}_\nu = e'_a{}^\mu \phi'^a{}_\nu + e'_a{}^\nu \phi'^a{}_\mu. \tag{13.42}$$

If we identify

$$e'_a{}^\rho \phi'^a{}_\mu =: \phi'^\rho{}_\mu, \tag{13.43}$$

we see that

$$\psi^{\rho}{}_{\mu} = \phi'^{\rho}{}_{\mu} + \phi'_{\mu}{}^{\rho}. \tag{13.44}$$

Even though $\phi'^{\rho}{}_{\mu}$ is not in principle symmetric, the perturbation in the metric—which is usually supposed to represent a fundamental spin-2 field—is to be identified with the symmetric part of $\phi'^{\rho}{}_{\mu}$. It is then easy to see that, in terms of $\psi^{\rho}{}_{\mu}$, the gauge transformation (13.19) acquires the form

$$\psi^{\rho}{}_{\mu} \to \psi^{\rho}{}_{\mu} - (1/\sqrt{k})\big[\partial^{\rho}\xi_{\mu}(x) + \partial_{\mu}\xi^{\rho}(x)\big], \tag{13.45}$$

where

$$\xi^{\mu}(x) = \xi^{a}(x) e'_{a}{}^{\mu}. \tag{13.46}$$

The Bianchi identity (13.21), on the other hand, is seen to be trivially satisfied, whereas the field equation (13.27) assumes the form

$$\Box\big(\delta^{\mu}{}_{\lambda}\psi - \psi^{\mu}{}_{\lambda}\big) - \partial_{\lambda}\partial^{\mu}\psi - \delta^{\mu}{}_{\lambda}\partial_{\nu}\partial_{\rho}\psi^{\nu\rho} + \partial_{\nu}\partial^{\mu}\psi^{\nu}{}_{\lambda} + \partial_{\lambda}\partial_{\rho}\psi^{\rho\mu} = 0, \tag{13.47}$$

with $\psi = \psi^{\alpha}{}_{\alpha}$. This is precisely the linearized Einstein equation, which means that, in absence of gravity, the teleparallel-based approach is totally equivalent to the usual General Relativity-based approach to the spin-2 field. Namely, $\phi^{a}{}_{\rho}$ and $\psi_{\mu\nu}$ are the same physical field. The teleparallel approach, however, is much more elegant and simple, in the sense that it is similar to the spin-1 electromagnetic theory—an heritage of the (abelian) gauge structure of teleparallel gravity. In addition, it allows a precise distinction between gauge transformations—local translations in the tangent space—and spacetime coordinate transformations.

13.4 Coupling to Gravitation

13.4.1 Gravitational Coupling Prescription

In absence of gravitation, as we have seen in the previous section, the algebraic index of $\phi^{a}{}_{\rho}$ can be transformed into a spacetime index through contraction with the tetrad field, and vice-versa. In the presence of gravitation, this index transformation can lead to problems with the coupling prescription. In fact, as already said, a spin-2 field theory presents consistency problems when coupled to gravitation [8, 9]. The problem is that the divergence identities satisfied by the field equations of a spin-2 field on Minkowski spacetime are no longer valid once the coupling is made. In addition, the coupled equations are no longer gauge invariant. The basic underlying difficulty is related to the fact that the covariant derivative of General Relativity—which defines the gravitational coupling prescription—is non-commutative, and this introduces unphysical constraints on the spacetime curvature. As we are going to see here, all such inconsistencies disappear if Teleparallel Gravity is used as paradigm, and a spin-2 field is represented, not by a symmetric rank-two tensor, but by a translational-valued 1-form.

To begin with we note that, because $\phi^a{}_\rho$ is a vector field assuming values in the Lie algebra of the translation group, $\phi_\rho = \phi^a{}_\rho P_a$, the algebraic index "$a$" is not an ordinary vector index. It is actually a gauge index which, due to the "external" character of translations, happens to be similar to the usual, true vector index "ρ". However, as a translational gauge index, it is irrelevant for the gravitational coupling prescription. This means that, in the class of frames h'_a in which the inertial connection $\dot{A}^a{}_{b\mu}$ vanishes, the gravitational coupling prescription of the spin-2 field $\phi_\rho = \phi^a{}_\rho P_a$ is written in the form [see Chap. 5]

$$\partial_\mu \phi'_\rho \to \partial_\mu \phi'_\rho - (\dot{\Gamma}^\lambda{}_{\rho\mu} - \dot{K}^\lambda{}_{\rho\mu})\phi'_\lambda. \tag{13.48}$$

Comment 13.2 Of course, because of the identity

$$\dot{\Gamma}^\lambda{}_{\rho\mu} - \dot{K}^\lambda{}_{\rho\mu} = \mathring{\Gamma}^\lambda{}_{\rho\mu}, \tag{13.49}$$

with $\mathring{\Gamma}^\lambda{}_{\rho\mu}$ the Levi-Civita connection of the metric $g_{\mu\nu}$, this coupling prescription coincides with the coupling prescription of General Relativity. This is a key point of the equivalence between General Relativity and Teleparallel Gravity [10].

In components, the coupling prescription (13.48) reads

$$\partial_\mu \phi'^a{}_\rho \to \partial_\mu \phi'^a{}_\rho - (\dot{\Gamma}^\lambda{}_{\rho\mu} - \dot{K}^\lambda{}_{\rho\mu})\phi'^a{}_\lambda. \tag{13.50}$$

In a Lorentz rotated class of frames

$$h_a = \Lambda_a{}^b(x) h'_b,$$

the gravitational coupling prescription assumes the form

$$\partial_\mu \phi^a{}_\rho \to \dot{\mathscr{D}}_\mu \phi^a{}_\rho - (\dot{\Gamma}^\lambda{}_{\rho\mu} - \dot{K}^\lambda{}_{\rho\mu})\phi^a{}_\lambda, \tag{13.51}$$

where

$$\dot{\mathscr{D}}_\mu \phi^a{}_\rho = \partial_\mu \phi^a{}_\rho + \dot{A}^a{}_{b\mu}\phi^b{}_\rho. \tag{13.52}$$

This prescription provides different couplings, in terms of connection-terms, for each index of $\phi^a{}_\rho$: whereas the algebraic index is connected to inertial effects only, the spacetime index is coupled to terms representing the gravitational effects. It constitutes one of the main differences between the teleparallel-based approach and the usual metric approach of General Relativity. In fact, the latter considers both indices of the spin-2 variable $\psi_{\mu\nu}$ on an equal footing, leading to a coupling prescription that breaks the gauge invariance of the spin-2 theory.

Comment 13.3 The consistency of this procedure can be verified from the very definitions of curvature and torsion. As is well-known, the curvature of a connection A is defined as *the covariant derivative of A in the connection A*. In the language of differential forms,

$$R = dA + A \wedge A. \tag{13.53}$$

Similarly, the torsion of a connection A is defined as *the covariant derivative of the tetrad h in the connection A*:

$$T = dh + A \wedge h. \tag{13.54}$$

In components, these expressions assume the form

$$R_{\mu\nu} = \partial_\mu A_\nu - \partial_\nu A_\mu + [A_\mu, A_\nu] \tag{13.55}$$

and

$$T_{\mu\nu} = \partial_\mu h_\nu - \partial_\nu h_\mu + [A_\mu, h_\nu]. \tag{13.56}$$

We see from these definitions that the algebraic indices are irrelevant for the covariant derivatives defining curvature and torsion. They have to do only with the inertial properties of the frame. The same holds for the gravitational coupling prescription (13.51).

13.4.2 Field Strength and Bianchi Identity

Let us now apply the gravitational coupling prescription (13.51) to the free theory. To begin with we notice that, because the gauge parameter ξ^a has an algebraic index only, the gauge transformation (13.34) does not change in the presence of gravitation:

$$\phi^a{}_\mu \to \phi^a{}_\mu - (1/\sqrt{k})\dot{\mathscr{D}}_\mu \xi^a(x). \tag{13.57}$$

On the other hand, considering that the connection (13.49) is symmetric in the last two indices, we see that neither the field strength $F^a{}_{\mu\nu}$, defined by Eq. (13.35), changes in the presence of gravitation:

$$F^a{}_{\mu\nu} = \dot{\mathscr{D}}_\mu \phi^a{}_\nu - \dot{\mathscr{D}}_\nu \phi^a{}_\mu. \tag{13.58}$$

Due to the fact that the teleparallel Fock-Ivanenko derivative $\dot{\mathscr{D}}_\mu$ is commutative, the Bianchi identity also remains unchanged,

$$\dot{\mathscr{D}}_\rho F^a{}_{\mu\nu} + \dot{\mathscr{D}}_\nu F^a{}_{\rho\mu} + \dot{\mathscr{D}}_\mu F^a{}_{\nu\rho} = 0. \tag{13.59}$$

In terms of the Levi-Civita *tensor* $\varepsilon^{\lambda\rho\mu\nu}$, it can be rewritten in the form

$$\dot{\mathscr{D}}_\rho \big(h \varepsilon^{\lambda\rho\mu\nu} F^a{}_{\mu\nu} \big) = 0. \tag{13.60}$$

This is similar to what happens to the electromagnetic field in the presence of gravitation.

13.4.3 Lagrangian and Field Equation

In a way analogous to the flat background case, the lagrangian of the spin-2 field in the presence of gravitation can be obtained from the teleparallel lagrangian (9.13) by replacing the teleparallel torsion $\dot{T}^a{}_{\mu\nu}$ by the spin-2 field strength $\sqrt{k} F^a{}_{\mu\nu}$. The result is

$$\mathscr{L} = \frac{h}{4} F^a{}_{\mu\nu} \mathscr{F}_a{}^{\mu\nu}, \tag{13.61}$$

where

$$\mathscr{F}_a{}^{\mu\nu} = h_a{}^{\rho} \mathscr{K}^{\mu\nu}{}_{\rho} - h_a{}^{\mu} h_b{}^{\sigma} F^{b\nu}{}_{\sigma} + h_a{}^{\nu} h_b{}^{\sigma} F^{b\mu}{}_{\sigma} \tag{13.62}$$

is the gravitationally-coupled field excitation, with

$$\mathscr{K}^{\mu\nu}{}_{\rho} = \tfrac{1}{2}\left(h_a{}^{\nu} F^{a\mu}{}_{\rho} + h^a{}_{\rho} F_a{}^{\mu\nu} - h_a{}^{\mu} F^{a\nu}{}_{\rho}\right) \tag{13.63}$$

the corresponding spin-2 analogous of the coupled contortion. We notice in passing that this lagrangian is invariant under the gauge transformation (13.34). It is furthermore invariant under local Lorentz transformation of the frames.

Performing variations in relation to $\phi^a{}_\rho$, we get

$$\dot{\mathscr{D}}_\mu \mathscr{F}_a{}^{\rho\mu} + \left(\dot{\Gamma}^{\mu}{}_{\nu\mu} - \dot{K}^{\mu}{}_{\nu\mu}\right) \mathscr{F}_a{}^{\rho\nu} = 0. \tag{13.64}$$

This expression, by using the identity

$$\partial_\mu h = h \mathring{\Gamma}^{\mu}{}_{\lambda\mu} \equiv h\left(\dot{\Gamma}^{\mu}{}_{\lambda\mu} - \dot{K}^{\mu}{}_{\lambda\mu}\right), \tag{13.65}$$

can be rewritten in the form

$$\dot{\mathscr{D}}_\mu \left(h \mathscr{F}_a{}^{\rho\mu}\right) = 0. \tag{13.66}$$

This is the field equation of a fundamental spin-2 field in the presence of gravitation, as seen from the general frame $h^a{}_\mu$. It can also be obtained directly from the free field equation (13.40) by applying the gravitational coupling prescription (13.51). The crucial property of this equation is that it is both gauge invariant and local Lorentz invariant.

Summing up: although in absence of gravitation the teleparallel-based approach to the spin-2 field coincides with the usual metric approach based on General Relativity [see Sect. 13.3.6], in the presence of gravitation it differs substantially. The reason is that the index "a" of the translational-valued field $\phi^a{}_\rho$ is not an ordinary vector index, but a gauge index. As such, it is irrelevant for the gravitational coupling prescription. In the metric approach both indices of the spin-2 field $\psi_{\mu\nu}$ are considered on an equal footing. The ensuing gravitational coupling prescription, which includes connection terms for both indices of $\psi_{\mu\nu}$, is found to break the gauge invariance of the theory. As a consequence, it is not possible to eliminate the spurious degrees of freedom in such a way as to remain with just two independent components. When the teleparallel coupling prescription is used, on the other hand, the gravitationally-coupled spin-2 field theory that emerges is both gauge and local-Lorentz invariant. It is quite similar to the gravitationally-coupled electromagnetic theory, and preserves the duality symmetry of the free theory.

13.5 Spin-2 Field as Source of Gravitation

Let us consider now the total lagrangian

$$\mathscr{L}_t = \dot{\mathscr{L}} + \mathscr{L}_s, \tag{13.67}$$

where $\overset{\bullet}{\mathcal{L}}$ is the teleparallel lagrangian (9.13), and $\mathcal{L}_s$ is the lagrangian (13.61) of a spin-2 field in the presence of gravitation. The corresponding field equation is

$$\partial_\sigma\big(h\overset{\bullet}{S}_a{}^{\rho\sigma}\big) - kh\big(\overset{\bullet}{t}_a{}^{\rho} + \overset{\bullet}{\imath}_a{}^{\rho}\big) = kh\Theta_a{}^{\rho}, \tag{13.68}$$

where $\overset{\bullet}{t}_a{}^{\rho}$ is the gravitational energy-momentum tensor, $\overset{\bullet}{\imath}_a{}^{\rho}$ is the energy-momentum pseudotensor of inertia, and

$$\Theta_a{}^{\rho} \equiv -\frac{1}{h}\frac{\delta\mathcal{L}_s}{\delta h^a{}_\rho} = h_a{}^{\nu}\mathcal{F}_c{}^{\mu\rho}F^c{}_{\mu\nu} - \frac{h_a{}^{\rho}}{h}\mathcal{L}_s \tag{13.69}$$

is the spin-2 field source energy-momentum tensor. Notice that

$$\Theta_\rho{}^{\rho} \equiv h^a{}_\rho\Theta_a{}^{\rho} = 0, \tag{13.70}$$

as it should be for a massless field. Furthermore, from the invariance of $\mathcal{L}_s$ under a general coordinate transformation, the energy-momentum tensor is found to satisfy the usual covariant conservation law

$$\overset{\circ}{\mathcal{D}}_\rho\big(h\Theta_a{}^{\rho}\big) \equiv \partial_\rho\big(h\Theta_a{}^{\rho}\big) - \overset{\circ}{A}{}^b{}_{a\rho}\big(h\Theta_b{}^{\rho}\big) = 0. \tag{13.71}$$

Due to the anti-symmetry of the superpotential in the last two indices, we see from the field equation (13.68) that the total energy-momentum density is conserved in the ordinary sense:

$$\partial_\rho\big[h\big(\overset{\bullet}{\imath}_a{}^{\rho} + \overset{\bullet}{t}_a{}^{\rho} + \Theta_a{}^{\rho}\big)\big] = 0. \tag{13.72}$$

As we have seen in Chap. 9, the field equation (13.68) can be rewritten as

$$\overset{\bullet}{\mathcal{D}}_\sigma\big(h\overset{\bullet}{S}_a{}^{\rho\sigma}\big) = kh\big(\overset{\bullet}{t}_a{}^{\rho} + \Theta_a{}^{\rho}\big), \tag{13.73}$$

where the right-hand side represents the true source of gravitation. We recall that the energy-momentum density coming from the interaction of inertial effects of the frame with gravity, although entering the total energy-momentum conservation, is not source of gravitation, and must accordingly remain in the left-hand side of the field equation. Considering that the covariant derivative $\overset{\bullet}{\mathcal{D}}_\rho$ is commutative [see Sect. 10.3], and taking into account the anti-symmetry of the superpotential in the spacetime indices, we obtain the divergence identity

$$\overset{\bullet}{\mathcal{D}}_\rho\overset{\bullet}{\mathcal{D}}_\sigma\big(h\overset{\bullet}{S}_a{}^{\rho\sigma}\big) = 0, \tag{13.74}$$

which implies that the true source of gravitation is conserved in the covariant sense:

$$\overset{\bullet}{\mathcal{D}}_\rho\big[h\big(\overset{\bullet}{t}_a{}^{\rho} + \Theta_a{}^{\rho}\big)\big] = 0. \tag{13.75}$$

These properties, together with the gauge and local Lorentz invariance, which reduce the number of degrees of freedom to only two, render the teleparallel-based gravitationally-coupled spin-2 theory fully consistent [11].

Comment 13.4 In the context of General Relativity, whose spin connection represents both inertia and gravitation, the inertial part of the gravitational energy-momentum pseudotensor cannot be separated, and consequently the gravitational field equation cannot be written in a form equivalent to (13.73). In this context, therefore, no consistent gravitationally-coupled spin-2 field theory can be obtained. This is similar to the definition of a tensorial energy-momentum density for the gravitational field, which cannot be given in the context of General Relativity.

13.6 On Gravitational Waves

Although equivalent to General Relativity, Teleparallel Gravity introduces conceptual changes into gravitation. One of these changes refers to the concept of gravitational waves. For massless particles belonging to a the Lorentz representation (A, B), the quantity

$$\sigma = B - A \tag{13.76}$$

denotes the helicity of the representation [12]. Since according to General Relativity gravitational waves are interpreted as curvature waves, whose excitation 2-form $\mathscr{R}^{\rho}{}_{\lambda\mu\nu}$ belongs to the representation

$$(0, 2) \oplus (2, 0) \tag{13.77}$$

of the Lorentz group, they are believed to be spin-2 waves with helicity $\sigma = \pm 2$. There is a problem, though: when coupled to gravitation, the field equation describing such waves become inconsistent [8, 9]. In fact, in addition to imposing unphysical constraints on spacetime, the lack of gauge invariance makes it impossible to eliminate the spurious degrees of freedom of the theory.

On the other hand, in the teleparallel approach gravitational waves are interpreted as waves of the field excitation 2-form $\mathscr{F}^{a\mu\nu}$, which is a translational-valued 2-form that belongs to the representation

$$(1/2, 3/2) \oplus (3/2, 1/2) \tag{13.78}$$

of the Lorentz group. As such, they describe spin-2 waves with helicity $\sigma = \pm 1$. From the point of view of representation theory, both representations $(0, 2) \oplus (2, 0)$ and $(1/2, 3/2) \oplus (3/2, 1/2)$ are well-defined Lorentz representations, and consequently able to describe physical waves. There is a difference, though: whereas in the first case the gravitationally-coupled equations are inconsistent, in the second they are fully consistent—as shown in the previous sections. This result seems to suggest that gravitational waves should be interpreted, not as curvature waves with helicity $\sigma = \pm 2$, but as a translational-valued field excitation waves with helicity $\sigma = \pm 1$.

It should be remarked that, since the concept of helicity can only be defined at the linear level [13], these arguments hold at this level only. Considering that, on account of the nonlinear character of gravitation, gravitational waves are essentially nonlinear [14, 15], it is not clear whether the concepts related to a linear spin-2 field can be immediately extended to gravitational waves. Even if the linear approximation to the gravitational waves theory makes physical sense, it could always be corrected by higher-order terms, in which case the very notion of helicity would be lost.

References

1. Birrel, N.D., Davies, P.C.W.: Quantum Fields in Curved Spacetime. Cambridge University Press, Cambridge (1982)

2. Deser, S., Henneaux, M.: Class. Quantum Gravity **24**, 1683 (2007). gr-qc/0611157
3. Dirac, P.A.M.: In: Kockel, B., Macke, W., Papapetrou, A. (eds.) Planck Festscrift. Deutscher Verlag der Wissenschaften, Berlin (1958)
4. Pauli, W., Fierz, M.: Helv. Phys. Acta **12**, 297 (1939)
5. Fierz, M., Pauli, W.: Proc. R. Soc. Lond. Ser. A **173**, 211 (1939)
6. Obukhov, Yu.N., Pereira, J.G.: Phys. Rev. D **67**, 044008 (2003). gr-qc/0212078
7. Gribl Lucas, T., Pereira, J.G.: J. Phys. A **42**, 035402 (2009). 0811.2066 [gr-qc]
8. Aragone, C., Deser, S.: Nuovo Cimento A **3**, 4 (1971)
9. Aragone, C., Deser, S.: Nuovo Cimento B **57**, 33 (1980)
10. de Andrade, V.C., Guillen, L.C.T., Pereira, J.G.: Phys. Rev. D **64**, 027502 (2001). gr-qc/0003100
11. Arcos, H.I., Gribl Lucas, T., Pereira, J.G.: Class. Quantum Gravity **27**, 145007 (2010). 1001.3407 [gr-qc]
12. Weinberg, S.: The Quantum Theory of Fields, vol. 1, p. 254. Cambridge University Press, Cambridge (1995)
13. Wald, R.M.: General Relativity. University of Chicago Press, Chicago (1984)
14. Aldrovandi, R., Pereira, J.G., Vu, K.H.: Found. Phys. **37**, 1503 (2007). 0709.1603 [gr-qc]
15. Aldrovandi, R., Pereira, J.G., da Rocha, R., Vu, K.H.: Int. J. Theor. Phys. **49**, 549 (2010). 0809.2911 [gr-qc]

Chapter 14
Teleparallel Equivalent of Some Solutions

General Relativity and Teleparallel Gravity are constructed on the very same metric spacetime—the same metric and tetrad field are at work in both approaches. They use, however, different connections: the torsionless Levi-Civita connection in the first, and the vanishing-curvature Weitzenböck connection in the second. For the sake of comparison with well-known results of General Relativity, the torsion tensor associated by Teleparallel Gravity to the de Sitter and the Kerr solutions are obtained.

14.1 de Sitter Spacetime

In spite of their equivalence, Teleparallel Gravity and General Relativity exhibit important conceptual differences. For example, General Relativity describes gravitation in terms of the curvature tensor, while Teleparallel Gravity describes it in terms of torsion. To illustrate this fact, we are going to study in this section the teleparallel version of the de Sitter solution.

14.1.1 de Sitter Torsion

The de Sitter spacetime, denoted $dS(4,1)$, is a four-dimensional, maximally-symmetric pseudo-riemannian space with constant sectional curvature [see Ref. [1] for an interesting account of this spacetime]. It is a homogeneous space under the Lorentz group $\mathscr{L} = SO(3,1)$,

$$dS(4,1) = SO(4,1)/\mathscr{L}, \tag{14.1}$$

with $SO(4,1)$ the de Sitter group [2]. In the context of General Relativity, it shows up as a solution to the sourceless Einstein equation with a cosmological term Λ:

$$\overset{\circ}{R}_{\mu\nu} - \tfrac{1}{2} g_{\mu\nu} \overset{\circ}{R} - g_{\mu\nu} \Lambda = 0. \tag{14.2}$$

R. Aldrovandi, J.G. Pereira, *Teleparallel Gravity*, Fundamental Theories of Physics 173,
DOI 10.1007/978-94-007-5143-9_14, © Springer Science+Business Media Dordrecht 2013

In stereographic coordinates [3], the de Sitter metric assumes the conformally flat form

$$g_{\mu\nu} = \Omega^2 \eta_{\mu\nu}, \tag{14.3}$$

with the conformal factor given by

$$\Omega \equiv \Omega(x) = \frac{1}{1 - (\sigma^2 \Lambda/12)}, \tag{14.4}$$

where σ^2 is the Lorentz-invariant quadratic interval

$$\sigma^2 = \eta_{\mu\nu} x^\mu x^\nu. \tag{14.5}$$

In the limit of a vanishing cosmological term, $g_{\mu\nu}$ reduces to the Minkowski metric:

$$\lim_{\Lambda \to 0} g_{\mu\nu} = \eta_{\mu\nu}. \tag{14.6}$$

The Christoffel connection of metric (14.3) is [4]

$$\overset{\circ}{\Gamma}{}^\rho{}_{\mu\nu} = \left(\delta^\rho_\mu \delta^\sigma_\nu + \delta^\rho_\nu \delta^\sigma_\mu - \eta_{\mu\nu} \eta^{\rho\sigma}\right) \partial_\sigma \ln \Omega, \tag{14.7}$$

with the ensuing Riemann tensor components given by

$$\overset{\circ}{R}{}^\mu{}_{\nu\rho\sigma} = -\frac{\Lambda}{3} \left(\delta^\mu_\rho g_{\nu\sigma} - \delta^\mu_\sigma g_{\nu\rho}\right). \tag{14.8}$$

The Ricci and the scalar curvature tensors are, consequently,

$$\overset{\circ}{R}_{\mu\nu} = -\Lambda g_{\mu\nu} \quad \text{and} \quad \overset{\circ}{R} = -4\Lambda. \tag{14.9}$$

14.1.2 The de Sitter Torsion

Let us now obtain the teleparallel equivalent of this solution. From the relation

$$g_{\mu\nu} = \eta_{ab} h^a{}_\mu h^b{}_\nu, \tag{14.10}$$

with $g_{\mu\nu}$ given by Eq. (14.3), the de Sitter tetrad is found to be [5]

$$h^a{}_\mu = \Omega \delta^a_\mu. \tag{14.11}$$

Considering the class of frames in which $\overset{\bullet}{A}{}^a{}_{b\mu} = 0$, the Weitzenböck connection [see Eq. (4.65)] assumes the form

$$\overset{\bullet}{\Gamma}{}^\rho{}_{\mu\nu} = h_a{}^\rho \partial_\nu h^a{}_\mu. \tag{14.12}$$

In the specific case of the de Sitter tetrad (14.11), it is

$$\overset{\bullet}{\Gamma}{}^\rho{}_{\mu\nu} = \delta^\rho_\mu \partial_\nu \ln \Omega. \tag{14.13}$$

The torsion tensor of the teleparallel equivalent for the de Sitter solution is, consequently,

$$\overset{\bullet}{T}{}^\rho{}_{\mu\nu} = \delta^\rho_\nu \partial_\mu \ln \Omega - \delta^\rho_\mu \partial_\nu \ln \Omega. \tag{14.14}$$

Using Eq. (14.4), it can be rewritten as

$$\dot{T}^{\rho}{}_{\mu\nu} = \frac{\Lambda}{6\Omega}(\delta^{\rho}_{\nu} g_{\mu\alpha} - \delta^{\rho}_{\mu} g_{\nu\alpha})x^{\alpha}. \tag{14.15}$$

The corresponding contortion tensor is

$$\dot{K}^{\rho}{}_{\mu\nu} = \frac{\Lambda}{6\Omega}(\delta^{\rho}_{\alpha} g_{\mu\nu} - \delta^{\rho}_{\nu} g_{\mu\alpha})x^{\alpha}. \tag{14.16}$$

The possibility of decomposing the torsion tensor into three parts which are irreducible under the group of global Lorentz transformations allows a better understanding of its physical meaning. As discussed in Sect. 1.7, that decomposition assumes the form

$$\dot{T}_{\lambda\mu\nu} = \tfrac{2}{3}(\dot{\mathcal{T}}_{\lambda\mu\nu} - \dot{\mathcal{T}}_{\lambda\nu\mu}) + \tfrac{1}{3}(g_{\lambda\mu}\dot{\mathcal{V}}_{\nu} - g_{\lambda\nu}\dot{\mathcal{V}}_{\mu}) + \varepsilon_{\lambda\mu\nu\rho}\dot{\mathcal{A}}^{\rho}, \tag{14.17}$$

where $\dot{\mathcal{T}}_{\lambda\mu\nu}$ is the purely tensor part, and $\dot{\mathcal{V}}_{\mu}$ and $\dot{\mathcal{A}}^{\rho}$ represent respectively the vector and axial parts of torsion. In the specific case of the de Sitter spacetime, only the vector torsion is non-vanishing, and has the form

$$\dot{\mathcal{V}}_{\rho} = -\frac{\Lambda}{2\Omega} g_{\rho\alpha} x^{\alpha}. \tag{14.18}$$

The homogeneity and isotropy of the de Sitter spacetime implies the vanishing of both the purely tensor and the axial-vector torsions.

14.1.3 The de Sitter Force

In Teleparallel Gravity, the force equation that replaces the geodesic equation of General Relativity is [see Eq. (6.53)]

$$\frac{dp^{\rho}}{ds} + \dot{\Gamma}^{\rho}{}_{\mu\nu} p^{\mu} u^{\nu} = F^{\rho}, \tag{14.19}$$

where $p^{\rho} = mcu^{\rho}$ is the particle four-momentum, and

$$F^{\rho} = mc\dot{K}^{\rho}{}_{\mu\nu} u^{\mu} u^{\nu} \tag{14.20}$$

represents the gravitational force. Substituting (14.16), we obtain

$$F^{\rho} = \frac{\Lambda mc}{6\Omega} P^{\rho}{}_{\nu} x^{\nu}, \tag{14.21}$$

where $P^{\rho}{}_{\nu} = \delta^{\rho}_{\nu} - u^{\rho} u_{\nu}$ is a velocity-orthogonal projection tensor. In terms of the vector torsion, on the other hand, it assumes the simple form

$$F^{\rho} = -\tfrac{1}{3} mc P^{\rho}{}_{\nu} \dot{\mathcal{V}}^{\nu}. \tag{14.22}$$

Let us now obtain the gravitational force in the non-relativistic limit [see Sect. 6.3 for a description of the general procedure]. As is well known, this limit is achieved by taking the velocity of light c going to infinity: $c \to \infty$. However, in the presence of Λ, this limit yields an acceptable contraction limit only if $\Lambda \to 0$, but in such a way that

$$\frac{3}{c^2\Lambda} \equiv \tau^2 \tag{14.23}$$

remains finite [6]. The reason for such condition is that, if c is taken to infinity keeping Λ fixed, the resulting theory would not have a ten-dimensional kinematic group. In particular, no boosts would be present, which is not a physically consistent result. The parameter τ can be interpreted as a kind of "Hubble time" [7]. Taking the limit with this condition in mind, the conformal factor turns out to be

$$\Omega = \frac{1}{1-(t^2/4\tau^2)}, \tag{14.24}$$

with the consequent gravitational force vector

$$\boldsymbol{F} = \frac{m\boldsymbol{r}}{2[1-(t^2/4\tau^2)]\tau^2}. \tag{14.25}$$

Since $t < \tau$, the force is always positive (or repulsive). In the limit of a vanishing cosmological term $\Lambda \to 0$, which corresponds to $\tau \to \infty$, we obtain a free particle.

14.2 Teleparallel Equivalent of the Kerr Solution

We are going to study now the teleparallel version of the Kerr solution.[1] When the source angular momentum vanishes, it reduces to the Schwarzschild solution. In Boyer-Lindquist coordinates (r, θ, φ), the Kerr metric is written as [8]

$$ds^2 = g_{00}\,dt^2 + g_{11}\,dr^2 + g_{22}\,d\theta^2 + g_{33}\,d\varphi^2 + 2g_{03}\,d\varphi\,dt, \tag{14.26}$$

where

$$g_{00} = 1 - \frac{r_S r}{\rho^2}, \qquad g_{11} = -\frac{\rho^2}{\Delta}, \qquad g_{22} = -\rho^2, \tag{14.27}$$

$$g_{33} = -\left(r^2 + a^2 + \frac{r_S r a^2}{\rho^2}\sin^2\theta\right)\sin^2\theta, \tag{14.28}$$

$$g_{03} = g_{30} = \frac{r_S r a}{\rho^2}\sin^2\theta. \tag{14.29}$$

In these expressions,

$$r_s = 2Gm \tag{14.30}$$

is the Schwarzschild radius, with m the mass of the source, and

$$\rho^2 = r^2 + a^2\cos^2\theta, \qquad \Delta = r^2 - r_S r + a^2, \tag{14.31}$$

with a the angular momentum of a unit mass. For $a = 0$ the Kerr metric reduces to the standard form of the Schwarzschild solution.

[1] In this section we use units with $c = 1$.

14.2.1 Kerr Torsion

Using the relation

$$g_{\mu\nu} = \eta_{ab} h^a{}_\mu h^b{}_\nu, \tag{14.32}$$

the components of the Kerr tetrad are found to be

$$h^a{}_\mu \equiv \begin{pmatrix} \gamma_{00} & 0 & 0 & \eta \\ 0 & \gamma_{11}\mathrm{s}\theta\mathrm{c}\varphi & \gamma_{22}\mathrm{c}\theta\mathrm{c}\varphi & -\beta\mathrm{s}\varphi \\ 0 & \gamma_{11}\mathrm{s}\theta\mathrm{s}\varphi & \gamma_{22}\mathrm{c}\theta\mathrm{s}\varphi & \beta\mathrm{c}\varphi \\ 0 & \gamma_{11}\mathrm{c}\theta & -\gamma_{22}\mathrm{s}\theta & 0 \end{pmatrix}, \tag{14.33}$$

where

$$\beta^2 = \eta^2 - g_{33} \quad \text{and} \quad \eta = g_{03}/\gamma_{00}. \tag{14.34}$$

Furthermore, the following notations have been introduced:

$$\gamma_{00} = \sqrt{g_{00}}, \qquad \gamma_{ii} = \sqrt{-g_{ii}}, \tag{14.35}$$

and

$$s\theta = \sin\theta, \qquad c\theta = \cos\theta, \qquad s\varphi = \sin\varphi, \qquad c\varphi = \cos\varphi. \tag{14.36}$$

The inverse tetrad is given by

$$h_a{}^\mu \equiv \begin{pmatrix} \gamma_{00}^{-1} & 0 & 0 & 0 \\ -\beta g^{03}\mathrm{s}\varphi & \gamma_{11}^{-1}\mathrm{s}\theta\mathrm{c}\varphi & \gamma_{22}^{-1}\mathrm{c}\theta\mathrm{c}\varphi & -\beta^{-1}\mathrm{s}\varphi \\ \beta g^{03}\mathrm{c}\varphi & \gamma_{11}^{-1}\mathrm{s}\theta\mathrm{s}\varphi & \gamma_{22}^{-1}\mathrm{c}\theta\mathrm{s}\varphi & \beta^{-1}\mathrm{c}\varphi \\ 0 & \gamma_{11}^{-1}\mathrm{c}\theta & -\gamma_{22}^{-1}\mathrm{s}\theta & 0 \end{pmatrix}. \tag{14.37}$$

For the sake of simplicity, we choose now to work in the class of frames in which the inertial spin connection vanishes:

$$\overset{\bullet}{A}{}^a{}_{b\mu} = 0. \tag{14.38}$$

In this case, the Weitzenböck connection (4.65) assumes the form

$$\overset{\bullet}{\Gamma}{}^\rho{}_{\nu\mu} = h_a{}^\rho \partial_\mu h^a{}_\nu. \tag{14.39}$$

Through a lengthy but straightforward calculation, their non-vanishing components are found to be

$$\dot{\Gamma}^0{}_{01} = [\ln\sqrt{g_{00}}\,]_{,r} \qquad \dot{\Gamma}^0{}_{13} = \beta g^{03}\gamma_{11}\mathrm{s}\theta$$

$$\dot{\Gamma}^0{}_{23} = -\beta g^{03}\gamma_{22}\mathrm{c}\theta \qquad \dot{\Gamma}^0{}_{31} = \eta_{,r}/\gamma_{00} - (\beta^2)_{,r}g^{03}/2$$

$$\dot{\Gamma}^0{}_{32} = \eta_{,\theta}/\gamma_{00} - (\beta^2)_{,\theta}g^{03}/2$$

$$\dot{\Gamma}^1{}_{11} = [\ln\sqrt{-g_{11}}\,]_{,r} \qquad \dot{\Gamma}^1{}_{12} = [\ln\sqrt{-g_{11}}\,]_{,\theta}$$

$$\dot{\Gamma}^1{}_{22} = -\gamma_{22}/\gamma_{11} \qquad \dot{\Gamma}^1{}_{33} = -\beta\mathrm{s}\theta/\gamma_{11}$$

$$\dot{\Gamma}^2{}_{12} = \gamma_{11}/\gamma_{22} \qquad \dot{\Gamma}^2{}_{21} = [\ln\sqrt{-g_{22}}\,]_{,r}$$

$$\dot{\Gamma}^2{}_{22} = [\ln\sqrt{-g_{22}}\,]_{,\theta} \qquad \dot{\Gamma}^2{}_{33} = -\beta\mathrm{c}\theta/\gamma_{22}$$

$$\dot{\Gamma}^3{}_{13} = \gamma_{11}\mathrm{s}\theta/\beta \qquad \dot{\Gamma}^3{}_{23} = \gamma_{22}\mathrm{c}\theta/\beta$$

$$\dot{\Gamma}^3{}_{31} = [\ln\beta]_{,r} \qquad \dot{\Gamma}^3{}_{32} = [\ln\beta]_{,\theta}$$

where ordinary derivatives have been denoted by a "comma". The corresponding non-zero components of the torsion tensor (4.70) are found to be [9]

$$\begin{aligned}
\dot{T}^0{}_{01} &= -(\ln\gamma_{00})_{,r}\\
\dot{T}^0{}_{13} &= \eta_{,r}/\gamma_{00} - \beta g^{03}(\beta_{,r} - \gamma_{11}\mathrm{s}\theta)\\
\dot{T}^0{}_{23} &= \eta_{,\theta}/\gamma_{00} - \beta g^{03}(\beta_{,\theta} - \gamma_{22}\mathrm{c}\theta)\\
\dot{T}^1{}_{12} &= -(\ln\gamma_{11})_{,\theta}\\
\dot{T}^2{}_{12} &= (\ln\gamma_{22})_{,r} - \gamma_{11}/\gamma_{22}\\
\dot{T}^3{}_{13} &= (\beta_{,r} - \gamma_{11}\mathrm{s}\theta)/\beta\\
\dot{T}^3{}_{23} &= (\beta_{,\theta} - \gamma_{22}\mathrm{c}\theta)/\beta.
\end{aligned}$$

For the Kerr solution, vector, axial-vector and purely tensorial parts of torsion are non-zero. In particular, the non-vanishing axial-vector components are

$$\dot{\mathscr{A}}^{(r)} = -\frac{1}{3h}\left[g_{00}\dot{T}^0{}_{23} + g_{03}\dot{T}^3{}_{23}\right] \tag{14.40}$$

and

$$\dot{\mathscr{A}}^{(\theta)} = \frac{1}{3h}\left[g_{00}\dot{T}^0{}_{13} + g_{03}\left(\dot{T}^3{}_{13} + \dot{T}^0{}_{01}\right)\right], \tag{14.41}$$

where, we recall, $h = \det(h^a{}_\mu)$. When the angular momentum a vanishes, the axial torsion vanishes as well, and both the vector and the purely tensorial parts of torsion reduce to the values of the Schwarzschild solution.

References

1. Schrödinger, E.: Expanding Universes. Cambridge University Press, Cambridge (1956)

2. Hawking, S.W., Ellis, G.F.R.: The Large Scale Structure of Space-Time. Cambridge University Press, Cambridge (1973)
3. Gürsey, F.: In: Gürsey, F. (ed.) Group Theoretical Concepts and Methods in Elementary Particle Physics. Istanbul Summer School of Theoretical Physics. Gordon and Breach, New York (1962)
4. Aldrovandi, R., Pereira, J.G.: An Introduction to Geometrical Physics. World Scientific, Singapore (1995)
5. Aldrovandi, R., Barbosa, A.L., Calçada, M., Pereira, J.G.: Found. Phys. **33**, 613 (2003). gr-qc/0105068
6. Bacry, H., Lévy-Leblond, J.M.: J. Math. Phys. **9**, 1605 (1968)
7. Gibbons, G.W., Patricot, C.E.: Class. Quantum Gravity **20**, 5225 (2003). gr-qc/0308200
8. Misner, C.W., Thorne, K.S., Wheeler, J.A.: Gravitation. Freeman, San Francisco (1973)
9. Pereira, J.G., Vargas, T., Zhang, C.M.: Class. Quantum Gravity **18**, 833 (2001). gr-qc/0102070

Chapter 15
Duality Symmetry

Duality symmetry is an important property of *internal* sourceless gauge theories. It says that the dynamic field equation is, in absence of source, just the Bianchi identity written for the dual of the field strength, and vice versa. This means that, if we know the geometry, we automatically know the dynamics. Duality is present neither in General Relativity nor in Teleparallel Gravity. Taking advantage of the possibility of separating torsion into irreducible components under the global Lorentz group, a dual-symmetric toy sub-theory of Teleparallel Gravity is obtained.

15.1 Introduction

General Relativity does not exhibit the remarkable symmetry of duality, a characteristic of *internal* gauge theories [see Sect. 3.1]. Teleparallel Gravity, on the other hand, is a gauge theory, though for the *external* translation group [see Chap. 4]. It presents several properties distinguishing it from General Relativity, and at the same time many others shared with internal gauge theories. For example, it does not describe the gravitational interaction by a geometrization of spacetime, as General Relativity does, but by a force [see Chap. 6]. Another important point refers to the lagrangian of the gravitational field: whereas in General Relativity it is linear in the field excitation, or curvature, in Teleparallel Gravity, similarly to internal gauge theories, the lagrangian is quadratic in the field strength of the theory—which is torsion. On what concerns the duality symmetry, Teleparallel Gravity shares with General Relativity the property of not exhibiting duality symmetry—a lack of symmetry which seems then to be an intrinsic characteristic of gravity.

The main difference that Teleparallel Gravity shows with respect to the internal gauge theories comes precisely from its external fingerprint: the presence of a solder form connecting the internal with the external sectors of the theory. This form, whose components are the tetrad field, gives rise to new types of contractions, absent in internal gauge theories. In consequence the gauge lagrangian, as well as the field equation, will include additional terms if compared to the internal theories. As discussed in Chap. 8, these additional terms can be taken into account through a

R. Aldrovandi, J.G. Pereira, *Teleparallel Gravity*, Fundamental Theories of Physics 173,
DOI 10.1007/978-94-007-5143-9_15,

soldered-bundle generalization of the concept of Hodge dual [1]. This new definition opens the door for the study of duality symmetry of gravity in the context of Teleparallel Gravity [2].

15.2 Duality Symmetry and Gravitation

Consider the first Bianchi identity of Teleparallel Gravity, as given by Eq. (9.74):

$$\dot{\mathscr{D}}_\nu \dot{T}^a{}_{\rho\mu} + \dot{\mathscr{D}}_\mu \dot{T}^a{}_{\nu\rho} + \dot{\mathscr{D}}_\rho \dot{T}^a{}_{\mu\nu} = 0. \tag{15.1}$$

It can equivalently be written in the form

$$\dot{\mathscr{D}}_\rho \big(\varepsilon^{\lambda\rho\mu\nu} \dot{T}^a{}_{\mu\nu}\big) = 0. \tag{15.2}$$

Written for the dual torsion, it reads

$$\dot{\mathscr{D}}_\rho \big(\varepsilon^{\lambda\rho\mu\nu} \star \dot{T}^a{}_{\mu\nu}\big) = 0. \tag{15.3}$$

Substituting the generalized Hodge dual [see Eq. (8.26)]

$$\star \dot{T}^a{}_{\mu\nu} = \tfrac{1}{2} h \varepsilon_{\mu\nu\alpha\beta} \dot{S}^{a\alpha\beta}, \tag{15.4}$$

and using the relation [see Eq. (1.92)],

$$\varepsilon^{\lambda\rho\mu\nu} \varepsilon_{\mu\nu\alpha\beta} = -2\big(\delta^\lambda_\alpha \delta^\rho_\beta - \delta^\lambda_\beta \delta^\rho_\alpha\big), \tag{15.5}$$

it reduces to

$$\dot{\mathscr{D}}_\sigma \big(h \dot{S}_a{}^{\rho\sigma}\big) = 0. \tag{15.6}$$

Comparing with the sourceless gravitational field equation [see Eq. (10.17)]

$$\dot{\mathscr{D}}_\sigma \big(h \dot{S}_a{}^{\rho\sigma}\big) - k\big(h \dot{t}_a{}^\rho\big) = 0, \tag{15.7}$$

we see that the Bianchi identity written for the dual torsion does not yield the sourceless gravitational field equation: it yields the potential term of the field equation, but not the current term. This means that gravitation is not dual-symmetric. The condition for gravitation to present duality symmetry, therefore, is that the purely gravitational (or *tensorial*) energy-momentum current, given by Eq. (10.18), vanishes identically:

$$\dot{t}_a{}^\rho \equiv \frac{1}{k} h_a{}^\lambda \dot{S}_c{}^{\nu\rho} \dot{T}^c{}_{\nu\lambda} - \frac{h_a{}^\rho}{h} \dot{\mathscr{L}} = 0. \tag{15.8}$$

Comment 15.1 As discussed in Sect. 10.5, the *covariant part* of the gauge self-current of the internal gauge theories vanishes identically. In other words, these currents have only a non-covariant piece, which is included in the potential term of the field equation. We see now that this is the very reason why Yang-Mills type gauge theories are dual symmetric.

Using the expression (9.13) for the teleparallel lagrangian, we arrive at the equivalent condition

$$\dot{S}_c{}^{\mu\rho} \dot{T}^c{}_{\mu\lambda} = \tfrac{1}{4} \delta^\rho_\lambda \dot{S}_c{}^{\mu\nu} \dot{T}^c{}_{\mu\nu}. \tag{15.9}$$

Although this is quite a restrictive constraint, the possibility is not excluded that under some specific conditions gravitation may present duality symmetry.

15.3 Linear Gravity

A trivial case in which condition (15.8) is fulfilled is linear gravity. In fact, considering that the energy-momentum current is at least quadratic in the field variable—as, by the way, any other current—it necessarily vanishes in the linear approximation. To see this explicitly, let us take the trivial (non-gravitational) tetrad

$$e^a{}_\mu \equiv \dot{\mathscr{D}}_\mu x^a = \partial_\mu x^a + \dot{A}^a{}_{b\mu} x^b, \tag{15.10}$$

discussed in Chap. 1. In the class of frames in which the inertial connection $\dot{A}^a{}_{b\mu}$ vanishes, it becomes

$$e^a{}_\mu = \partial_\mu x^a. \tag{15.11}$$

For such tetrads, it is always possible to properly choose the translational gauge and the coordinate system in such a way that

$$e^a{}_\mu = \delta^a_\mu. \tag{15.12}$$

In this case, the spacetime metric assumes the usual form

$$\eta_{\mu\nu} = \delta^a_\mu \delta^b_\nu \eta_{ab} = \text{diag}(+1, -1, -1, -1). \tag{15.13}$$

Next, we consider the gauge potential $B^a{}_\mu$ as a small perturbation in relation to the trivial tetrad, and write

$$B^a{}_\mu = \varepsilon B^a_{(1)\mu} + \varepsilon^2 B^a_{(2)\mu} + \cdots, \tag{15.14}$$

where ε is a small dimensionless parameter introduced to label the successive orders of perturbation. We obtain, in consequence, an expansion of the tetrad field around the flat background:

$$h^a{}_\mu = \delta^a_\mu + \varepsilon B^a_{(1)\mu} + \varepsilon^2 B^a_{(2)\mu} + \cdots. \tag{15.15}$$

The corresponding expansion of the metric tensor is

$$g_{\mu\nu} = \eta_{\mu\nu} + \varepsilon (B_{(1)\mu\nu} + B_{(1)\nu\mu}) + \cdots, \tag{15.16}$$

where we have introduced the notation

$$B^\rho_{(1)\nu} = \delta^\rho_a B^a_{(1)\nu}. \tag{15.17}$$

The spacetime indices are now raised and lowered with the Minkowski metric, as in

$$B_{(1)\mu\nu} = \eta_{\mu\rho} B^\rho_{(1)\nu}. \tag{15.18}$$

Although the perturbation $B_{(1)\mu\nu}$ of the gauge potential is not in principle symmetric, it has already been shown that the anti-symmetric part of $B_{(1)\mu\nu}$ drops off completely from the first-order lagrangian and field equation [3]. For this reason, we are going to assume from now on that $B_{(1)\mu\nu}$ is symmetric. In this case, and up to first order, the Weitzenböck connection (4.65) for $\dot{A}^a{}_{b\mu} = 0$ reads

$$\dot{\Gamma}^\rho_{(1)\mu\nu} = \partial_\nu B^\rho_{(1)\mu}. \tag{15.19}$$

The torsion and the contortion tensors are found to be, respectively,

$$\dot{T}^{\rho}_{(1)\mu\nu} = \partial_\mu B^{\rho}_{(1)\nu} - \partial_\nu B^{\rho}_{(1)\mu} \tag{15.20}$$

and

$$\dot{K}^{\rho}_{(1)\mu\nu} = \partial^\rho B_{(1)\mu\nu} - \partial_\mu B^{\rho}_{(1)\nu}. \tag{15.21}$$

The corresponding first-order superpotential is

$$\begin{aligned}\dot{S}_{(1)\nu}{}^{\rho\mu} = {} & \partial^\rho B^{\mu}_{(1)\nu} - \partial^\mu B^{\rho}_{(1)\nu} - \delta^{\mu}_{\nu}\big(\partial^\rho B^{\sigma}_{(1)\sigma} - \partial_\sigma B^{\sigma\rho}_{(1)}\big) \\ & + \delta^{\rho}_{\nu}\big(\partial^\mu B^{\sigma}_{(1)\sigma} - \partial_\sigma B^{\sigma\mu}_{(1)}\big).\end{aligned} \tag{15.22}$$

The corresponding first-order sourceless gravitational field equation is then

$$\partial_\mu \dot{S}_{(1)a}{}^{\rho\mu} = 0. \tag{15.23}$$

We point out once more that the superpotential (15.22) coincides formally with the Fierz tensor [4, 5] for a symmetric spin-2 field [see Sect. 13.3.3].

From Eq. (15.6) we see that the first-order version of the Bianchi identity written for the dual torsion is

$$\partial_\mu \dot{S}_{(1)\nu}{}^{\rho\mu} = 0, \tag{15.24}$$

where we have identified

$$\delta^{a}_{\nu} \dot{S}_{(1)a}{}^{\rho\mu} = \dot{S}_{(1)\nu}{}^{\rho\mu}. \tag{15.25}$$

Comparing the first-order field equation (15.23) with the first-order Bianchi identity (15.24) written for the dual torsion, we conclude that linear gravity does present duality symmetry. This is actually an expected result because, as already remarked in Chap. 12, linear General Relativity presents duality symmetry [6] and Teleparallel Gravity is equivalent to General Relativity. It is important to remark, however, that the generalized Hodge dual (15.4) must be used to get this result.

Another important implication of this generalized dual refers to the action functional of the gravitational field which, according to Eq. (9.4), is given by

$$\dot{\mathscr{S}}[\dot{T}] = \frac{c^3}{16\pi G} \int \eta_{ab} \dot{T}^a \wedge \star \dot{T}^b. \tag{15.26}$$

Written for the dual, it reads

$$\dot{\mathscr{S}}[\star \dot{T}] = \frac{c^3}{16\pi G} \int \eta_{ab} \star \dot{T}^a \wedge \star\star \dot{T}^b. \tag{15.27}$$

Since

$$\star\star \dot{T}^b = - \dot{T}^b,$$

we see immediately that the action changes sign under the dual transformation:

$$\dot{\mathscr{S}}[\star \dot{T}] = - \dot{\mathscr{S}}[\dot{T}]. \tag{15.28}$$

This is a known property of linear General Relativity [6], which emerges quite trivially in the context of Teleparallel Gravity. Furthermore, it is found to hold not only for the linear case, but also for the full theory.

15.4 Looking for a Dual Gravity

As discussed in Sect. 12.4.2, the gravitational interaction of a Dirac spinor ψ in Teleparallel Gravity involves only the vector and the axial parts of torsion [7]. In fact, in the class of frames in which the inertial connection vanishes, the teleparallel Dirac equation reads

$$i\hbar\gamma^{\mu}\big(\partial_{\mu} - \tfrac{1}{2}\dot{\mathcal{V}}_{\mu} - \tfrac{3i}{4}\dot{\mathcal{A}}_{\mu}\gamma^{5}\big)\psi = mc\psi. \tag{15.29}$$

Then comes the point: considering that the pure tensor piece $\dot{\mathcal{T}}_{\lambda\mu\nu}$ is irrelevant for the description of the gravitational interaction of spinor fields, if we restrict ourselves to the microscopic world of the fermions, it is sensible to consider a gravitational sub-theory in which the pure tensor piece of torsion is absent. In this case, torsion reduces to

$$\dot{T}_{\lambda\mu\nu} = \tfrac{1}{3}\big(g_{\lambda\mu}\dot{\mathcal{V}}_{\nu} - g_{\lambda\nu}\dot{\mathcal{V}}_{\mu}\big) + \varepsilon_{\lambda\mu\nu\rho}\dot{\mathcal{A}}^{\rho}. \tag{15.30}$$

The corresponding contortion tensor is

$$\dot{K}^{\rho\mu\nu} = \tfrac{1}{3}\big(g^{\nu\rho}\dot{\mathcal{V}}^{\mu} - g^{\nu\mu}\dot{\mathcal{V}}^{\rho}\big) - \tfrac{1}{2}\varepsilon^{\nu\rho\mu\lambda}\dot{\mathcal{A}}_{\lambda}, \tag{15.31}$$

whereas the superpotential becomes

$$\dot{S}^{\rho\mu\nu} = -\tfrac{2}{3}\big(g^{\rho\mu}\dot{\mathcal{V}}^{\nu} - g^{\rho\nu}\dot{\mathcal{V}}^{\mu}\big) - \tfrac{1}{2}\varepsilon^{\rho\mu\nu\lambda}\dot{\mathcal{A}}_{\lambda}. \tag{15.32}$$

Substituting these expressions in the teleparallel lagrangian

$$\dot{\mathcal{L}} = \frac{c^4 h}{32\pi G}\dot{T}_{\rho\mu\nu}\dot{S}^{\rho\mu\nu}, \tag{15.33}$$

we obtain

$$\dot{\mathcal{L}} = \frac{c^4 h}{16\pi G}\big(-\tfrac{2}{3}\dot{\mathcal{V}}_{\mu}\dot{\mathcal{V}}^{\mu} + \tfrac{3}{2}\dot{\mathcal{A}}_{\mu}\dot{\mathcal{A}}^{\mu}\big). \tag{15.34}$$

Comment 15.2 An alternative way to get this lagrangian is to note that the teleparallel lagrangian (15.33) can be written in the form [8]

$$\dot{\mathcal{L}} = \frac{c^4 h}{16\pi G}\big(\tfrac{2}{3}\dot{\mathcal{T}}^{\lambda\mu\nu}\dot{\mathcal{T}}_{\lambda\mu\nu} - \tfrac{2}{3}\dot{\mathcal{V}}^{\mu}\dot{\mathcal{V}}_{\mu} + \tfrac{3}{2}\dot{\mathcal{A}}^{\mu}\dot{\mathcal{A}}_{\mu}\big). \tag{15.35}$$

When the purely tensor part $\dot{\mathcal{T}}^{\lambda\mu\nu}$ vanishes, it just reduces to (15.34).

Would this theory present duality symmetry? To answer this question, we rewrite the condition for a gravitational theory to present duality symmetry—given by Eq. (15.9)—in terms of the vector and the axial torsions. The result is

$$\begin{aligned}-\tfrac{4}{9}\dot{\mathcal{V}}_{\lambda}\dot{\mathcal{V}}^{\rho} - \tfrac{2}{9}\delta_{\lambda}{}^{\rho}\dot{\mathcal{V}}_{\mu}\dot{\mathcal{V}}^{\mu} - \tfrac{1}{2}\varepsilon^{\rho}{}_{\lambda\nu\alpha}\dot{\mathcal{V}}^{\nu}\dot{\mathcal{A}}^{\alpha} + \delta_{\lambda}^{\rho}\dot{\mathcal{A}}_{\mu}\dot{\mathcal{A}}^{\mu} \\ - \dot{\mathcal{A}}_{\lambda}\dot{\mathcal{A}}^{\rho} = \tfrac{1}{2}\delta_{\lambda}^{\rho}\big(-\tfrac{2}{3}\dot{\mathcal{V}}_{\mu}\dot{\mathcal{V}}^{\mu} + \tfrac{3}{2}\dot{\mathcal{A}}_{\mu}\dot{\mathcal{A}}^{\mu}\big).\end{aligned} \tag{15.36}$$

Now, as a simple inspection shows, no solution exists if torsion is real. However, in the complex domain, if the axial and vector parts of torsion are related by

$$\dot{\mathcal{A}}_{\mu} = \pm\tfrac{2i}{3}\dot{\mathcal{V}}_{\mu}, \tag{15.37}$$

the above condition is fulfilled, and the resulting gravitational theory turns out to present duality symmetry.

Applying the generalized dual definition (8.22) to the axial and vector torsions, we obtain

$$\star\dot{\mathscr{A}}_\mu = -\tfrac{2}{3}\dot{\mathscr{V}}_\mu \quad \text{and} \quad \star\dot{\mathscr{V}}_\mu = \tfrac{3}{2}\dot{\mathscr{A}}_\mu. \tag{15.38}$$

On account of the relation (15.37), we see that in this theory torsion turns out to be self-dual (upper sign) or anti-self-dual (lower sign):

$$\star\dot{\mathscr{A}}_\mu = \pm i\dot{\mathscr{A}}_\mu \quad \text{and} \quad \star\dot{\mathscr{V}}_\mu = \pm i\dot{\mathscr{V}}_\mu. \tag{15.39}$$

Notice that we are using here the extended notions of self-duality and anti-self-duality for complex fields [9]. In that context we can say that there exists a sub-theory of Teleparallel Gravity which is able to describe the gravitational interaction of fermions and presents duality symmetry.

When the vector and axial-vector parts of torsion are related by Eq. (15.37), the Dirac equation (15.29) assumes the form

$$i\hbar\gamma^\mu\left[\partial_\mu - \tfrac{1}{2}\dot{\mathscr{V}}_\mu(1 \mp \gamma^5)\right]\psi = mc\psi, \tag{15.40}$$

with the upper (lower) sign referring to the self-dual (anti-self-dual) case. We see from this equation that a self-dual (anti-self-dual) torsion couples only to the left-hand (right-hand) component of the spinor field. In other words, at the microscopic level of the fermions, gravitation becomes a chiral interaction. Observe that the Dirac equation (15.40) is invariant under a chiral transformation

$$\psi \to \gamma^5\psi,$$

except for a change of sign of the spinor mass term. A similar property holds in Electrodynamics [see, for example, Ref. [10], p. 520].

15.5 Some Properties of the Self-dual Gravity

Let us briefly explore some properties of the self-dual gravitational theory obtained in the previous section. We begin by noting that, according to Eq. (15.37), torsion becomes a complex tensor. In fact, in terms of the vector torsion, it reads

$$\dot{T}_{\lambda\mu\nu} = \tfrac{1}{3}\left(g_{\lambda\mu}\dot{\mathscr{V}}_\nu - g_{\lambda\nu}\dot{\mathscr{V}}_\mu\right) \pm \tfrac{2i}{3}\varepsilon_{\lambda\mu\nu\rho}\dot{\mathscr{V}}^\rho. \tag{15.41}$$

The corresponding contortion tensor is

$$\dot{K}^{\rho\mu\nu} = \tfrac{1}{3}\left(g^{\nu\rho}\dot{\mathscr{V}}^\mu - g^{\nu\mu}\dot{\mathscr{V}}^\rho\right) \mp \tfrac{i}{3}\varepsilon^{\nu\rho\mu\lambda}\dot{\mathscr{V}}_\lambda, \tag{15.42}$$

whereas the superpotential acquires the form

$$\dot{S}^{\lambda\rho\sigma} = -\tfrac{2}{3}\left(g^{\lambda\rho}\dot{\mathscr{V}}^\sigma - g^{\lambda\sigma}\dot{\mathscr{V}}^\rho\right) \mp \tfrac{i}{3}\varepsilon^{\lambda\rho\sigma\theta}\dot{\mathscr{V}}_\theta. \tag{15.43}$$

Even though the torsion is complex, the gravitational lagrangian is found to be real:

$$\overset{\bullet}{\mathscr{L}} = -\frac{c^4 h}{12\pi G} \overset{\bullet}{\mathscr{V}}_\mu \overset{\bullet}{\mathscr{V}}^\mu. \tag{15.44}$$

The corresponding field equation is

$$\partial_\sigma \left(h \overset{\bullet}{S}_a{}^{\rho\sigma} \right) = 0. \tag{15.45}$$

Of course, as the theory is dual symmetric, this field equation coincides with the Bianchi identity written for the dual torsion, given by Eq. (15.6). Using the relation

$$\overset{\bullet}{S}_a{}^{\rho\sigma} = h_{a\lambda} \overset{\bullet}{S}{}^{\lambda\rho\sigma},$$

with $\overset{\bullet}{S}{}^{\lambda\rho\sigma}$ given by Eq. (15.43), its explicit form is the complex equation

$$-\tfrac{2}{3} \partial_\sigma \left[h \left(h_a{}^\rho \overset{\bullet}{\mathscr{V}}{}^\sigma - h_a{}^\sigma \overset{\bullet}{\mathscr{V}}{}^\rho \right) \right] + \tfrac{i}{3} \partial_\sigma \left(h h_{a\nu} \varepsilon^{\nu\rho\sigma\lambda} \overset{\bullet}{\mathscr{V}}_\lambda \right) = 0. \tag{15.46}$$

The imaginary part is

$$\partial_\sigma \left(h h_{a\nu} \varepsilon^{\nu\rho\sigma\lambda} \overset{\bullet}{\mathscr{V}}_\lambda \right) = 0, \tag{15.47}$$

which can be written in the equivalent form

$$\partial_\sigma \left(h h_{a\nu} \overset{\bullet}{\mathscr{V}}_\lambda \right) + \partial_\lambda \left(h h_{a\sigma} \overset{\bullet}{\mathscr{V}}_\nu \right) + \partial_\nu \left(h h_{a\lambda} \overset{\bullet}{\mathscr{V}}_\sigma \right) = 0. \tag{15.48}$$

This is the Bianchi identity of the theory. The real part of Eq. (15.46), on the other hand, is

$$-\tfrac{2}{3} \partial_\sigma \left[h \left(h_a{}^\rho \overset{\bullet}{\mathscr{V}}{}^\sigma - h_a{}^\sigma \overset{\bullet}{\mathscr{V}}{}^\rho \right) \right] = 0, \tag{15.49}$$

which is the dynamical field equation of the theory. We see then that, whereas the Bianchi identity shows up as the imaginary part, the dynamical field equation is obtained as the real part of the complex field equation (15.46). A similar mechanism holds in the Palatini formulation of self-dual General Relativity [see Ref. [11], p. 1511].

In the presence of a source field represented by $\mathscr{L}_s$, since the corresponding energy-momentum tensor

$$\Theta_a{}^\rho \equiv -\frac{1}{h} \frac{\delta \mathscr{L}_s}{\delta B^a{}_\rho} = -\frac{1}{h} \frac{\delta \mathscr{L}_s}{\delta h^a{}_\rho} \tag{15.50}$$

is real, it only contributes to the dynamical equation (15.49), which acquires then the form

$$\partial_\sigma \left[h \left(h_a{}^\sigma \overset{\bullet}{\mathscr{V}}{}^\rho - h_a{}^\rho \overset{\bullet}{\mathscr{V}}{}^\sigma \right) \right] = \frac{12\pi G}{c^4} h \Theta_a{}^\rho. \tag{15.51}$$

This is the equation governing the dynamics of the self-dual gravitational field. It should be remarked that, although this self-dual gravitational theory is able to describe the gravitational interaction of fermions, its physical meaning is still quite obscure. In particular, since its energy-momentum current vanishes, it seems unable to transport energy and momentum. Anyway, because it presents duality symmetry, this sub-theory has a natural intrinsic interest and may deserve further analysis.

Comment 15.3 In the weak field limit of (macroscopic) Teleparallel Gravity, the axial torsion is found not to contribute to the newtonian potential. In fact, only the vector and purely tensor parts of torsion contribute to Newton's potential [12]. On the other hand, in the (microscopic) limit of the gravitational interaction of fermions, it is the purely tensor part of torsion that does not contribute. As a consequence, the dual gravity theory will not have a newtonian limit. Of course, this is not a problem as this theory might be valid only at the microscopic level, where the newtonian limit is not required to hold.

Even though it has no meaning at the microscopic level of the gravitational interaction of fermions, it is instructive to obtain the (classical) equation of motion of a spinless particle in the presence of the dual gravitational field. In the context of Teleparallel Gravity, this equation of motion is given by [see Eq. (6.51)]

$$\frac{du^a}{ds} + \overset{\bullet}{A}{}^a{}_{b\rho} u^b u^\rho = \overset{\bullet}{K}{}^a{}_{b\rho} u^b u^\rho. \tag{15.52}$$

Substituting $\overset{\bullet}{K}{}^a{}_{b\rho}$ as given by Eq. (15.42), the imaginary part of torsion drops out, and we get

$$\frac{du^a}{ds} + \overset{\bullet}{A}{}^a{}_{b\rho} u^b u^\rho = -\tfrac{1}{3}(h^a{}_\rho - u^a u_\rho)\overset{\bullet}{\mathcal{V}}{}^\rho. \tag{15.53}$$

We see from this equation that the gravitational force in this case is real and has the form of a projector, being consequently orthogonal to the particle four-velocity, as it should be.

Comment 15.4 It is interesting to notice that the gravitational force (15.53) associated with the self-dual model, coincides with the gravitational force (14.22) associated with the de Sitter solution.

References

1. Gribl Lucas, T., Pereira, J.G.: J. Phys. A **42**, 035402 (2009). 0811.2066 [gr-qc]
2. de Andrade, V.C., Barbosa, A.L., Pereira, J.G.: Int. J. Mod. Phys. D **14**, 1635 (2005). gr-qc/0501037
3. Obukhov, Yu.N., Pereira, J.G.: Phys. Rev. D **67**, 044008 (2003). gr-qc/0212078
4. Pauli, W., Fierz, M.: Helv. Phys. Acta **12**, 297 (1939)
5. Fierz, M., Pauli, W.: Proc. R. Soc. Lond. Ser. A **173**, 211 (1939)
6. Henneaux, M., Teitelboim, C.: Phys. Rev. D **71**, 024018 (2005). gr-qc/0408101
7. Mosna, R.A., Pereira, J.G.: Gen. Relativ. Gravit. **36**, 2525 (2004). gr-qc/0312093
8. Hayashi, K., Shirafuji, T.: Phys. Rev. D **19**, 3524 (1979)
9. Ashtekar, A.: Class. Quantum Gravity **21**, R5 (2004)
10. Weinberg, S.: The Quantum Theory of Fields, vol. 1. Cambridge University Press, Cambridge (1995)
11. Ashtekar, A., Balachandran, A.P., Jo, S.: Int. J. Mod. Phys. A **4**, 1493 (1989)
12. Pereira, J.G., Vargas, T., Zhang, C.M.: Class. Quantum Gravity **18**, 833 (2001). gr-qc/0102070

Chapter 16
Teleparallel Kaluza-Klein Theory

In gauge theories the total (or bundle) space is *locally* a direct product of spacetime and the internal space of gauge variables. In unified models of the Kaluza-Klein type the bundle is taken as a *global* direct product. The Einstein-Hilbert lagrangian written on the total space leads, when the coordinates are conveniently separated, to the four-dimensional Einstein-Hilbert lagrangian plus a lagrangian of the gauge type. Replacing the geometric paradigm of General Relativity by the gauge paradigm of Teleparallel Gravity leads to new versions of Kaluza-Klein models, in which the unifying lagrangian is of the gauge type.

16.1 Introduction

In ordinary Kaluza-Klein theories [1], the geometrical approach of General Relativity is used as the paradigm for the description of all other interactions of Nature. In the original Kaluza-Klein theory, for example, gravitational and electromagnetic fields are described by a Einstein-Hilbert type lagrangian in a five-dimensional spacetime. On the other hand, the equivalence of Teleparallel Gravity, a gauge theory for the translation group, to General Relativity opens up a new road for the study of unified theories. In fact, instead of using the general-relativistic geometrical description, we can adopt the gauge description as the basic paradigm, and in this way construct what we call the *teleparallel equivalent of Kaluza-Klein theory* [2]. According to this approach, instead of a Einstein-Hilbert type lagrangian, the unifying lagrangian is of the gauge (or Maxwell) type. Before going through the details of this construction, we first review the basics of ordinary Kaluza-Klein theory.

16.2 Kaluza-Klein Theory: A Brief Review

In its simplest form, the Kaluza-Klein theory [3, 4] is an extension of General Relativity to a five dimensional pseudo-riemannian spacetime $\mathbb{R}^5$ with a topology given by the product between the usual four dimensional riemannian spacetime $\mathbb{R}^4$ and the circumference $\mathbb{S}^1$:

$$\mathbb{R}^5 = \mathbb{R}^4 \otimes \mathbb{S}^1.$$

R. Aldrovandi, J.G. Pereira, *Teleparallel Gravity*, Fundamental Theories of Physics 173,
DOI 10.1007/978-94-007-5143-9_16, © Springer Science+Business Media Dordrecht 2013

Denoting the five dimensional indices by capital Latin letters ($A, B, C, \ldots = 0$, 1, 2, 3, 5), the metric γ_{AB} of $\mathbb{R}^5$ is, in principle, a function of the coordinates x^μ of $\mathbb{R}^4$ and of the coordinates x^5 of $\mathbb{S}^1$:

$$\gamma_{AB} = \gamma_{AB}\big(x^\mu, x^5\big). \tag{16.1}$$

Since x^5 is a periodic coordinate, we can write

$$x^5 = \rho\theta, \tag{16.2}$$

with θ the angular coordinate and ρ the radius of $\mathbb{S}^1$. Using the metric γ_{AB}, we define now the five dimensional Levi-Civita connection [5]

$$\Gamma^C{}_{AB} = \tfrac{1}{2}\gamma^{CD}(\partial_A \gamma_{DB} + \partial_B \gamma_{DA} - \partial_D \gamma_{AB}). \tag{16.3}$$

Its curvature tensor is

$$R^A{}_{BCD} = \partial_C \Gamma^A{}_{BD} - \partial_D \Gamma^A{}_{BC} + \Gamma^A{}_{EC}\Gamma^E{}_{BD} - \Gamma^A{}_{ED}\Gamma^E{}_{BC}. \tag{16.4}$$

The action integral of the theory is written as a five dimensional Einstein-Hilbert lagrangian

$$\mathcal{S}_5 = -\frac{c^3}{16\pi G_5}\int d^5x\sqrt{-\gamma}\,R, \tag{16.5}$$

where $\gamma = \det(\gamma_{AB})$,

$$R = \gamma^{BD} R^A{}_{BAD} \tag{16.6}$$

is the five dimensional scalar curvature, and G_5 is the five-dimensional version of Newton's constant.

Comment 16.1 Around 1912, G. Nordström developed a scalar theory for gravitation [6–8]. He was the first to use a five-dimensional spacetime in an attempt to unify gravitation and electromagnetism [9].

The fifteen components of the five dimensional metric γ_{AB} can be represented by the four dimensional spacetime metric $g_{\mu\nu}$, a vector field A_μ, and a scalar field ϕ. In terms of these variables, it can be conveniently parametrized in the form

$$\gamma_{AB} = \left(\begin{array}{c|c} g_{\mu\nu} + \beta^2\phi A_\mu A_\nu & \beta\phi A_\mu \\ \hline \beta\phi A_\nu & \phi \end{array}\right), \tag{16.7}$$

with β a parameter to be determined in the unification procedure. On the other hand, the assumed topology for the five dimensional space allows us to expand any field quantity, and in particular each component of the metric γ_{AB}, in a Fourier series of the form

$$\gamma_{AB} = \sum_{n=-\infty}^{\infty} \gamma_{AB}^{(n)}(x^\mu)\exp\big(inx^5/r\big). \tag{16.8}$$

In order to obtain the four dimensional theory, Kaluza originally imposed the so-called *cylindric condition*

$$\frac{\partial \gamma_{AB}}{\partial x^5} = 0, \tag{16.9}$$

which corresponds to taking only the $n = 0$ mode in the Fourier expansion (16.8). Substituting the $n = 0$ piece of γ_{AB} into the action (16.5), integrating over x^5 and choosing

$$\beta^2 = \frac{16\pi G}{c^4}, \tag{16.10}$$

with

$$G = \frac{G_5}{2\pi\rho} \tag{16.11}$$

the ordinary Newton gravitational constant, we get [dropping the label $^{(0)}$ of the Fourier expansions]

$$\mathscr{S} = \int d^4x\sqrt{-g}\left(-\frac{c^3}{16\pi G}\mathring{R} + \frac{\phi}{4}F_{\mu\nu}F^{\mu\nu} + \frac{c^3}{24\pi G}\frac{\partial_\mu\phi\partial^\mu\phi}{\phi^2}\right), \tag{16.12}$$

where $g = \det(g_{\mu\nu})$, $\mathring{R}$ is the Ricci scalar curvature of the four dimensional space-time, and

$$F_{\mu\nu} = \partial_\mu A_\nu - \partial_\nu A_\mu. \tag{16.13}$$

If we take $\phi = -1$, we can then identify A_μ as the electromagnetic potential, and action (16.12) reduces to the Einstein-Maxwell action

$$\mathscr{S} = \int d^4x\sqrt{-g}\left(-\frac{c^3}{16\pi G}\mathring{R} - \frac{1}{4}F_{\mu\nu}F^{\mu\nu}\right), \tag{16.14}$$

with $F_{\mu\nu}$ the electromagnetic field strength.

This is usually considered the "miracle" of the Kaluza-Klein theories: gauge theories emerge naturally from geometry in the dimensional reduction process. It is particularly interesting to observe how the action (16.5), which is invariant under five dimensional general coordinate transformations, reduces to the action (16.12), which is invariant under both four dimensional general coordinate transformations and $U(1)$ gauge transformations. It is important to remark that, in order to get the proper relative sign between Einstein and Maxwell lagrangians, so that energy is positive, it is necessary that

$$\phi \equiv \gamma_{55} < 0. \tag{16.15}$$

According to our metric convention, this means that the fifth dimension must be space-like. This is consistent with causality, as more than one time-like dimension could lead to the existence of closed time-like curves.

A crucial property of the abelian Kaluza-Klein theory is that the five dimensional space $\mathbb{R}^5$ is a solution of the five dimensional Einstein equation

$$R_{AB} - \tfrac{1}{2}\gamma_{AB}R = 0, \tag{16.16}$$

which is obtained from action (16.5) through the variational principle. Using the decomposition (16.7), it yields respectively for $AB = \mu\nu$, $AB = \mu 5$ and $AB = 55$, the following equations

$$\mathring{R}_{\mu\nu} - \frac{1}{2}g_{\mu\nu}\mathring{R} = -\frac{\beta^2\phi}{2}\Theta_{\mu\nu} - \frac{1}{\phi}(\mathring{\nabla}_\mu(\partial_\nu\phi) - g_{\mu\nu}\Box\phi), \tag{16.17}$$

$$\overset{\circ}{\nabla}_\nu F_\mu{}^\nu = -3\frac{\partial^\nu \phi}{\phi} F_{\mu\nu}, \tag{16.18}$$

and

$$\Box\phi = -\frac{\beta^2\phi}{4} F_{\mu\nu}F^{\mu\nu}, \tag{16.19}$$

with $\overset{\circ}{R}_{\mu\nu}$ the four dimensional Ricci tensor, and $\Theta_{\mu\nu}$ the symmetric energy-momentum tensor of the electromagnetic field. If we set again $\phi = -1$, and use Eq. (16.10), we obtain the Einstein-Maxwell system of equations,

$$\overset{\circ}{R}_{\mu\nu} - \frac{1}{2} g_{\mu\nu} \overset{\circ}{R} = \frac{8\pi G}{c^4} \Theta_{\mu\nu} \tag{16.20}$$

and

$$\overset{\circ}{\nabla}_\nu F_\mu{}^\nu = 0. \tag{16.21}$$

The five dimensional *sourceless* field equation (16.16), therefore, reduces to the usual four dimensional Einstein equation with the electromagnetic energy-momentum tensor as source. This means that matter in four dimensions comes from the geometry of a five dimensional spacetime.

Condition $\phi = constant \neq 0$, used by Kaluza in his original paper, leads actually to inconsistencies. It is obvious from Eq. (16.19) that it implies

$$F_{\mu\nu}F^{\mu\nu} = 0, \tag{16.22}$$

as first noticed by Jordan [10], Bergmann [11] and Thiry [12]. This problem can be solved by taking into account all harmonics of the Fourier expansion (16.8). In this case, however, the scalar field ϕ will remain in the theory. Initially, this fact was considered a drawback of the theory because it leads actually to a scalar-tensor theory for gravitation. Afterwards, because scalar fields are useful to explain several phenomena of modern physics, like for example inflation and spontaneous symmetry breaking, this point became well accepted. Furthermore, quantum corrections would provide a mass for ϕ, thereby removing its long range gravitational effects [13]. Since these fields remain in the theory, they imply the existence of a huge family of new particles which emerge as a byproduct of the unification process.

Comment 16.2 The generalization of the Kaluza-Klein theory to non-abelian groups [14] can be made by assuming a spacetime manifold of the form

$$\mathbb{R}^{4+n} = \mathbb{R}^4 \otimes \mathbb{S}^n,$$

with $\mathbb{S}^n$ a compact manifold. In analogy with the five dimensional case, the dimensional reduction of a $(4+n)$-dimensional Einstein lagrangian yields four-dimensional Einstein's plus a Yang-Mills type lagrangian. It should be remarked however that, differently from the five dimensional abelian theory, the non-abelian case is plagued by conceptual difficulties. One of the main problems is that the space $\mathbb{R}^{4+n}$ cannot be a solution of the $(4+n)$-dimensional Einstein equation [15].

16.3 Five-Vector Potential

In the framework of Teleparallel Gravity, the action describing a particle of mass m and charge q in the presence of both an electromagnetic field A_μ and a gravitational field $B^a{}_\mu$ is [see Sect. 6.2.2]

$$\mathscr{S} = -mc \int_{\mathrm{p}}^{\mathrm{q}} \left(u_a \overset{\bullet}{\mathscr{D}}_\mu x^a + B^a{}_\mu u^b \eta_{ab} + \frac{q}{mc^2} A_\mu \right) dx^\mu. \tag{16.23}$$

The corresponding equation of motion is

$$h^a{}_\rho \frac{\overset{\bullet}{\mathscr{D}} u_a}{\mathscr{D}s} = \overset{\bullet}{T}{}^a{}_{\rho\mu} u^b u^\mu \eta_{ab} + \frac{q}{mc^2} F_{\rho\mu} u^\mu, \tag{16.24}$$

where

$$\overset{\bullet}{T}{}^a{}_{\mu\nu} = \overset{\bullet}{\mathscr{D}}_\mu B^a{}_\nu - \overset{\bullet}{\mathscr{D}}_\nu B^a{}_\mu \tag{16.25}$$

is the gravitational field strength—that is, torsion—and

$$F_{\mu\nu} = \overset{\bullet}{\mathscr{D}}_\mu A_\nu - \overset{\bullet}{\mathscr{D}}_\nu A_\mu \tag{16.26}$$

is the electromagnetic field strength, with $\overset{\bullet}{\mathscr{D}}_\mu A_\nu = \partial_\mu A_\nu$.

From the equation of motion (16.24) we see that torsion acts on particles in the very same way the electromagnetic field acts on electric charges, that is, as a force. This similarity allows a unified description of the gravitational and electromagnetic interactions. In order to get such a description, we begin by choosing the $U(1)$ gauge index of the electromagnetic theory as the *fifth* component of the gauge potential. Accordingly, we define a five-vector gauge potential in the form $(A, B, \ldots = 0, 1, 2, 3, 5)$

$$\mathscr{A}^A{}_\mu = \left(B^a{}_\mu, A^5{}_\mu\right), \tag{16.27}$$

where

$$A^5{}_\mu = \frac{q}{\kappa mc^2} A_\mu, \tag{16.28}$$

with κ a dimensionless parameter to be determined in the unification procedure. The unified field strength is consequently

$$\mathscr{F}^A{}_{\mu\nu} = \left(\overset{\bullet}{T}{}^a{}_{\mu\nu}, F^5{}_{\mu\nu}\right), \tag{16.29}$$

with

$$F^5{}_{\mu\nu} = \frac{q}{\kappa mc^2} F_{\mu\nu}. \tag{16.30}$$

In terms of the potential $\mathscr{A}^A{}_\mu$, therefore, it reads

$$\mathscr{F}^A{}_{\mu\nu} = \overset{\bullet}{\mathscr{D}}_\mu \mathscr{A}^A{}_\nu - \overset{\bullet}{\mathscr{D}}_\nu \mathscr{A}^A{}_\mu. \tag{16.31}$$

Implicit in the above definitions is the introduction of a five-dimensional space $\mathbb{M}^5$, the fiber of the model, given by the cartesian product between the Minkowski space $\mathbb{M}^4$ and the circle $\mathbb{S}^1$:

$$\mathbb{M}^5 = \mathbb{M}^4 \otimes \mathbb{S}^1.$$

A point in this space is determined by the coordinates

$$x^A = (x^a, x^5), \tag{16.32}$$

where x^a are the coordinates of $\mathbb{M}^4$, and x^5 is a coordinate on $\mathbb{S}^1$. The corresponding metric tensor is

$$\eta_{AB} = \begin{pmatrix} \eta_{ab} & 0 \\ 0 & \eta_{55} \end{pmatrix}. \tag{16.33}$$

We introduce now a velocity five-vector

$$u^A = (u^a, u^5). \tag{16.34}$$

The components u^a form the usual anholonomic four-velocity, and u^5 is a strictly internal component of u^A. In this case, by choosing

$$u^5 = -\kappa \quad \text{and} \quad \eta_{55} = -1, \tag{16.35}$$

action (16.23) can be rewritten in the form

$$\mathscr{S} = -mc \int_{\mathsf{p}}^{\mathsf{q}} \left(u_a \overset{\bullet}{\mathscr{D}}_\mu x^a + \mathscr{A}^A{}_\mu u^B \eta_{AB} \right) dx^\mu. \tag{16.36}$$

The corresponding equation of motion is

$$h^a{}_\rho \frac{\overset{\bullet}{\mathscr{D}} u_a}{\mathscr{D}s} = \mathscr{F}^A{}_{\rho\mu} u^B u^\mu \eta_{AB}. \tag{16.37}$$

Due to the fact that torsion acts on particles in the very same way the electromagnetic field acts on electric charges, the trajectory of a charged particle submitted to both electromagnetic and gravitational fields can be described by a unified Lorentz-type force equation.

Comment 16.3 Alternatively, we could have chosen

$$u^5 = \kappa \quad \text{and} \quad \eta_{55} = 1, \tag{16.38}$$

which would lead to the same action integral, and consequently to the same equation of motion. This choice corresponds to another metric convention for the internal space. In principle, both conventions are possible. However, as we are going to see in Sect. 16.5, the unification process will introduce a constraint according to which the choice of η_{55} will depend on the metric convention adopted for the tangent Minkowski space.

16.4 Teleparallel Kaluza-Klein

In a gauge theory for the translation group, the gauge transformation is defined as a local translation of the tangent-space coordinates,

$$\delta x^a = \varepsilon^b P_b x^a, \tag{16.39}$$

with $P_b = \partial/\partial x^b$ the generators, and ε^b the corresponding infinitesimal parameters. In a unified teleparallel Kaluza-Klein model, a general gauge transformation is represented by a translation of the five-dimensional space coordinates x^A,

$$\delta x^A = \varepsilon^B P_B x^A, \tag{16.40}$$

where $P_B = \partial/\partial x^B$ are the generators, and

$$\varepsilon^B = \left(\varepsilon^a, \varepsilon^5\right) \tag{16.41}$$

are the transformation parameters. Analogously to the definitions used for the gauge potentials, we define

$$\varepsilon^5 = \frac{q}{\kappa m c^2}\varepsilon. \tag{16.42}$$

Furthermore, in the same way as in ordinary Kaluza-Klein models, we assume that the gauge potentials $\mathscr{A}^A{}_\mu$, and consequently the tetrad $h^a{}_\mu$ and the metric tensor $g_{\mu\nu}$, do not depend on the coordinate x^5.

As seen in Chap. 9, the gravitational action of Teleparallel Gravity is

$$\overset{\bullet}{\mathscr{S}} = \frac{c^3}{16\pi G}\int \eta_{ab}\overset{\bullet}{T}{}^a \wedge \star\overset{\bullet}{T}{}^b, \tag{16.43}$$

with

$$\overset{\bullet}{T}{}^a = \tfrac{1}{2}\overset{\bullet}{T}{}^a{}_{\mu\nu}\, dx^\mu \wedge dx^\nu \tag{16.44}$$

the torsion 2-form. The action of the electromagnetic field, on the other hand, is

$$\mathscr{S}_{em} = -\tfrac{1}{4}\int F \wedge \star F, \tag{16.45}$$

with

$$F = \tfrac{1}{2}F_{\mu\nu}\, dx^\mu \wedge dx^\nu \tag{16.46}$$

the electromagnetic field 2-form. Owing to the gauge structure of both actions, it is possible to write down a unified action in the form

$$\mathscr{S} = \frac{c^3}{16\pi G}\int \eta_{AB}\mathscr{F}^A \wedge \star\mathscr{F}^B, \tag{16.47}$$

with

$$\mathscr{F}^A = \tfrac{1}{2}\mathscr{F}^A{}_{\mu\nu}\, dx^\mu \wedge dx^\nu \tag{16.48}$$

the unified 2-form field strength, whose components are given by Eq. (16.31).

Separating the spacetime and the electromagnetic components, the corresponding lagrangian is found to be

$$\mathscr{L} = \frac{hc^4}{32\pi G}\overset{\bullet}{T}{}^\rho{}_{\mu\nu}\overset{\bullet}{S}{}_\rho{}^{\mu\nu} + \eta_{55}\frac{\kappa^{-2}q^2}{16\pi G m^2}\frac{h}{4}F_{\mu\nu}F^{\mu\nu}. \tag{16.49}$$

The first term of $\mathscr{L}$ is the teleparallel gauge lagrangian (9.13), which is equivalent to the Einstein-Hilbert lagrangian of General Relativity. In order to get Maxwell's lagrangian from the second term, two conditions must be satisfied. First, it is necessary that

$$\kappa^2 = \frac{q^2}{16\pi G m^2}, \tag{16.50}$$

from where we see that κ^2 is the ratio between electric and gravitational forces. Second, in order to have a positive-definite energy for the electromagnetic field, and to get the appropriate relative sign between the gravitational and electromagnetic lagrangians, it is necessary that $\eta_{55} = -1$. With these conditions, the lagrangian (16.49) becomes

$$\mathcal{L} \equiv \overset{\bullet}{\mathcal{L}} + \mathcal{L}_{em} = \frac{h}{4k} \overset{\bullet}{T}{}^{\rho}{}_{\mu\nu} \overset{\bullet}{S}{}_{\rho}{}^{\mu\nu} - \frac{h}{4} F_{\mu\nu} F^{\mu\nu}, \tag{16.51}$$

with $k = 8\pi G/c^4$. As said in Sect. 16.2, that the Maxwell lagrangian in four dimensions shows up from the Einstein-Hilbert lagrangian in five dimensions, is usually considered as a miracle of the standard Kaluza-Klein theory [15]. That the Einstein-Hilbert lagrangian of General Relativity emerges from a gauge-type lagrangian for a five-dimensional translational gauge theory can be considered as another face of the same miracle.

The functional variation of $\mathcal{L}$ with respect to $B^a{}_\rho$ yields the teleparallel field equation

$$\partial_\sigma \big(h \overset{\bullet}{S}{}_a{}^{\rho\sigma} \big) - k \big(h \overset{\bullet}{J}{}_a{}^{\rho} \big) = k \big(h \Theta_a{}^{\rho} \big), \tag{16.52}$$

where $\overset{\bullet}{J}{}_a{}^{\rho}$ is the Noether energy-momentum current of the gravitational field, and

$$h \Theta_a{}^{\rho} \equiv - \frac{\delta \mathcal{L}_{em}}{\delta B^a{}_\rho} \equiv - \frac{\delta \mathcal{L}_{em}}{\delta h^a{}_\rho} = h F_{a\nu} F^{\rho\nu} - h_a{}^{\rho} \mathcal{L}_{em} \tag{16.53}$$

is the energy-momentum tensor of the electromagnetic field. On the other hand, the functional variation of $\mathcal{L}$ with respect to A_μ yields the teleparallel version of Maxwell's equation

$$\overset{\bullet\bullet}{\nabla}_\mu F^{\mu\nu} = 0, \tag{16.54}$$

as obtained in Sect. 12.5.

16.5 Metric Constraint

An interesting feature of the teleparallel Kaluza-Klein model is that it imposes a constraint between the spacetime metric and the metric on the $U(1)$ group. In fact, when we choose the metric of the Minkowski space to be

$$\eta_{ab} = \mathrm{diag}(+1, -1, -1, -1), \tag{16.55}$$

we have necessarily that $\eta_{55} = -1$, and the resulting metric of the five-dimensional space will be

$$\eta_{AB} = \mathrm{diag}(+1, -1, -1, -1, -1). \tag{16.56}$$

This means that the fifth dimension must necessarily be space-like, and the metric with signature $(3,2)$ is excluded. On the other hand, if we had chosen instead

$$\eta_{ab} = \text{diag}(-1,+1,+1,+1) \tag{16.57}$$

for Minkowski spacetime, it is easy to verify that the same consistency arguments would require that $\eta_{55} = 1$. The resulting metric of the five-dimensional space would then be

$$\eta_{AB} = \text{diag}(-1,+1,+1,+1,+1), \tag{16.58}$$

and the same conclusion would be obtained: the fifth dimension must necessarily be space-like, and the metric with signature $(3,2)$ is excluded. The unification of the gravitational and electromagnetic lagrangians, therefore, imposes a constraint on the metric conventions for Minkowski and for the electromagnetic internal manifold $\mathbb{S}^1$. In fact, the choice between $\eta_{55} = +1$ and $\eta_{55} = -1$ for the metric of $\mathbb{S}^1$ depends on the metric convention adopted for the Minkowski space. As a consequence, the metric of the five-dimensional internal space turns out to be restricted to either (16.56) or (16.58). Metrics with signature $(3,2)$, which would imply a time-like fifth dimension, are excluded.

16.6 Matter Fields

Let us consider now a general matter field Ψ. In contrast to the gauge fields, it depends on the coordinate x^5:

$$\Psi\left(x^A\right) = \Psi\left(x^\mu, x^5\right). \tag{16.59}$$

Under a generalized infinitesimal gauge translation (16.40), it transforms according to

$$\delta\Psi = \varepsilon^A P_A \Psi. \tag{16.60}$$

Considering that the internal manifold $\mathbb{S}^1$ is compact, we assume that the dependence of $\Psi(x^\mu, x^5)$ on the coordinate x^5 is of the form

$$\Psi\left(x^\mu, x^5\right) = \exp(i2\pi\theta)\psi\left(x^\mu\right), \tag{16.61}$$

where $\psi(x^\mu)$ depends on the spacetime coordinates only, and

$$\theta = \frac{\kappa x^5}{\lambda_C} \tag{16.62}$$

is the dimensionless coordinate (angle) of $\mathbb{S}^1$, with $\lambda_C = (2\pi\hbar/mc)$ the Compton wavelength of the particle represented by the wave function Ψ. That is to say,

$$\Psi\left(x^\mu, x^5\right) = \exp\left(i2\pi\frac{\kappa x^5}{\lambda_C}\right)\psi\left(x^\mu\right). \tag{16.63}$$

A translation in the coordinates x^a turns out to be a gauge transformation of the translation group:

$$\delta_a \Psi = \varepsilon^a \partial_a \Psi. \tag{16.64}$$

On the other hand, using that

$$\frac{\partial \Psi}{\partial x^5} = i 2\pi \frac{\kappa}{\lambda_C} \Psi, \tag{16.65}$$

a translation in the coordinate x^5 is found to be

$$\delta_5 \Psi = \varepsilon \left(\frac{iq}{\hbar c} \right) \Psi, \tag{16.66}$$

where we have used the identification (16.42). A translation in the internal coordinate x^5, therefore, turns out to be a $U(1)$ gauge transformation [see Sect. 11.2]. For a simultaneous infinitesimal translation in the five coordinates x^A, we see from transformation (16.60) that

$$\delta \Psi = \varepsilon^a \partial_a \Psi + \varepsilon \left(\frac{iq}{\hbar c} \right) \Psi. \tag{16.67}$$

In a general class of frames, the covariant derivative of Ψ is

$$\mathscr{D}_\mu \Psi = (\dot{\mathscr{D}}_\mu x^a) \partial_a \Psi + \mathscr{A}^A{}_\mu P_A \Psi, \tag{16.68}$$

with $\dot{\mathscr{D}}_\mu x^a$ given by (4.48). In the same class of frames, the infinitesimal gauge transformation of $\mathscr{A}^A{}_\mu$ is given by

$$\delta \mathscr{A}^B{}_\mu = -\dot{\mathscr{D}}_\mu \varepsilon^B. \tag{16.69}$$

Separating the five-potential $\mathscr{A}^A{}_\mu$ as given by Eq. (16.27), the covariant derivative (16.68) assumes the form

$$\mathscr{D}_\mu \Psi = h_\mu \Psi + \frac{iq}{\hbar c} A_\mu \Psi, \tag{16.70}$$

where $h_\mu = h^a{}_\mu \partial_a$, with

$$h^a{}_\mu = \dot{\mathscr{D}}_\mu x^a + B^a{}_\mu. \tag{16.71}$$

As usual, the commutator of covariant derivatives yields the field strength

$$[\mathscr{D}_\mu, \mathscr{D}_\nu] \Psi = \mathscr{F}^A{}_{\mu\nu} P_A \Psi, \tag{16.72}$$

with

$$\mathscr{F}^A{}_{\mu\nu} P_A = \dot{T}^a{}_{\mu\nu} P_a + \frac{iq}{\hbar c} F_{\mu\nu} \tag{16.73}$$

the field strength of the unified theory.

Comment 16.4 It is curious to observe that the field strength of gauge theories in general, and in particular the electromagnetic field strength $F_{\mu\nu}$, are interpreted as the *curvature* of the gauge potential A_μ. In other words, they represent the curvature of the internal space. Notwithstanding, in the teleparallel equivalent of the Kaluza-Klein model, the electromagnetic field strength appears as additional components of torsion, not of curvature. See Comment 9.6 for a related discussion on the Bianchi identities.

16.7 Further Remarks

The teleparallel equivalent of the standard Kaluza-Klein theory is a five-dimensional Maxwell-type translational gauge theory on a four-dimensional spacetime. In this theory, owing to the fact that both torsion and the electromagnetic field act on particles through a Lorentz type force, the electromagnetic field strength can be considered as an extra, fifth component of torsion. For this reason, the unification in this approach can be considered to be more natural than in the ordinary Kaluza-Klein theory.

An important feature of this model is that, in contrast to ordinary Kaluza-Klein models, no scalar field is generated by the unification process. Accordingly, no unphysical constraints appear, and the gravitational action can be naturally truncated at the zeroth mode. In other words, the cylindric condition can be naturally imposed for matter fields, which corresponds to keep only the $n = 0$ Fourier mode. The infinite spectrum of massive new particles is eliminated, strongly reducing the redundancy present in ordinary Kaluza-Klein theories. Furthermore, spacetime is kept four-dimensional, with no extra dimension. Only the "internal" space has additional dimensions.

Another important point concerns the relation between geometry and gauge theories. According to ordinary Kaluza-Klein models, gauge theories emerge from higher-dimensional geometric theories as a consequence of the dimensional reduction process. According to the teleparallel approach, however, gauge theories are the natural structures to be introduced, the four-dimensional geometry (gravitation) emerging from the non-compact sector of the "internal" space. In fact, only this sector can give rise to a tetrad field, which is the responsible for the geometrical structure induced on spacetime. As the gauge theories are introduced in their original form—they do not come from geometry—the unification turns out to be much more natural and easier to be performed.

The generalization of the teleparallel Kaluza-Klein model to non-abelian gauge theories is straightforward, and can be realized by introducing a $(4+n)$-dimensional internal space formed by the cartesian product between Minkowski space and a compact riemannian manifold, where n is the dimension of the gauge group [16]. Like in the electromagnetic case, the gauge field-strength appears as extra gauge components of torsion. It is worth mentioning also that most of the conceptual problems present in ordinary Kaluza-Klein models do not appear in the teleparallel version. In particular, due to the fact that the gauge structure—and not geometry—forms the basic paradigm, the problem discussed in Comment 16.2 does not exist in the non-abelian teleparallel Kaluza-Klein theory.

References

1. Appelquist, T., Chodos, A., Freund, P.G.O.: Modern Kaluza-Klein Theories. Addison-Wesley, Reading (1987)

2. de Andrade, V.C., Guillen, L.C.T., Pereira, J.G.: Phys. Rev. D **61**, 084031 (2000). gr-qc/9909004
3. Kaluza, Th.: Sitzungsber. Preuss. Akad. Wiss. Phys. Math. Kl. 966 (1921)
4. Klein, O.: Z. Phys. **37**, 895 (1926)
5. Overduin, J.M., Wesson, P.S.: Phys. Rep. **283**, 303 (1997)
6. Nordström, G.: Phys. Z. **13**, 1126 (1912)
7. Nordström, G.: Ann. Phys. (Leipz.) **40**, 872 (1913)
8. Nordström, G.: Ann. Phys. (Leipz.) **42**, 533 (1913)
9. Nordström, G.: Phys. Z. **15**, 504 (1914)
10. Jordan, P.: Ann. Phys. (Leipz.) **1**, 219 (1947)
11. Bergmann, P.: Ann. Math. **49**, 255 (1948)
12. Thiry, Y.: C. R. Acad. Sci. Paris **226**, 216 (1948)
13. Witten, E.: Nucl. Phys. B **186**, 412 (1981)
14. Kerner, R.: Ann. Inst. Henri Poincaré **9**, 143 (1968)
15. Chodos, A.: Comments Nucl. Part. Phys. **13**, 171 (1984)
16. Barbosa, A.L., Guillen, L.C.T., Pereira, J.G.: Phys. Rev. D **66**, 064028 (2002). gr-qc/0208052

Chapter 17
Einstein-Cartan Theory

In Teleparallel Gravity, torsion and curvature are related to the same degrees of freedom. There are theories, however, in which curvature and torsion represent different gravitational degrees of freedom. Of these, the Einstein-Cartan model can be considered the prototype. For the sake of comparison with Teleparallel Gravity, a brief review of this theory is presented. Some of its difficulties are pointed out.

17.1 Introduction

Some alternative gravitational models, like for example the Einstein-Cartan theory [1, 2] consider curvature and torsion as representing independent gravitational degrees of freedom. Torsion appears as intimately related to spin, and consequently only turns out to be important at the microscopic level, when spins become relevant. In this theory torsion is a non-propagating field. With independent works by Sciama [3] and Kibble [4] in the early nineteen-sixties, the original Einstein-Cartan theory was further developed in the form of a gauge theory for the Poincaré group, in which torsion becomes a propagating field. This generalization is known as Einstein-Cartan-Sciama-Kibble theory, in reference to those who have most contributed to Cartan's generalization of Einstein's theory. Considering that Einstein-Cartan is recovered as a particular case of the Poincaré gauge theory, and that these models present all the same relationship between torsion and spin, the Einstein-Cartan model can be taken as representative of this class, and for this reason it will be the only one to be discussed here.

The basic motivation for the Einstein-Cartan construction is the fact that, at a microscopic level, matter is represented by elementary particles, which are characterized by mass *and* spin. If one adopts the same *geometrical spirit of General Relativity*, not only mass but also spin should be source of gravitation at that level [for a textbook reference, see Ref. [5]]. In this line of thought, energy-momentum should keep its General Relativity role of source of curvature, whereas spin should appear as source of torsion. The relevant connection of this theory, therefore, is a general Cartan connection $A^a{}_{b\mu}$ with non-vanishing curvature and torsion:

R. Aldrovandi, J.G. Pereira, *Teleparallel Gravity*, Fundamental Theories of Physics 173, 171
DOI 10.1007/978-94-007-5143-9_17,

$$R^a{}_{b\nu\mu} \equiv \partial_\nu A^a{}_{b\mu} - \partial_\mu A^a{}_{b\nu} + A^a{}_{e\nu} A^e{}_{b\mu} - A^a{}_{e\mu} A^e{}_{b\nu} \neq 0 \tag{17.1}$$

and

$$T^a{}_{\nu\mu} \equiv \partial_\nu h^a{}_\mu - \partial_\mu h^a{}_\nu + A^a{}_{e\nu} h^e{}_\mu - A^a{}_{e\mu} h^e{}_\nu \neq 0. \tag{17.2}$$

As discussed in Chap. 1, that connection can be decomposed as [see Eq. (1.61)]

$$A^a{}_{b\mu} = \mathring{A}^a{}_{b\mu} + K^a{}_{b\mu}, \tag{17.3}$$

with $\mathring{A}^a{}_{b\mu}$ the spin connection of General Relativity. The corresponding decomposition in terms of the spacetime-indexed linear connection

$$\Gamma^\rho{}_{\nu\mu} = h_a{}^\rho \partial_\mu h^a{}_\nu + h_a{}^\rho A^a{}_{b\mu} h^b{}_\nu \equiv h_a{}^\rho \mathscr{D}_\mu h^a{}_\nu \tag{17.4}$$

is given by

$$\Gamma^\rho{}_{\nu\mu} = \mathring{\Gamma}^\rho{}_{\nu\mu} + K^\rho{}_{\nu\mu}, \tag{17.5}$$

with $\mathring{\Gamma}^\rho{}_{\nu\mu}$ the Levi-Civita connection.

17.2 Field Equations

The Einstein-Cartan gravitational lagrangian is

$$\mathscr{L}_{EC} = -\frac{c^4}{16\pi G}\sqrt{-g}R. \tag{17.6}$$

Although it formally coincides with the Einstein-Hilbert lagrangian of General Relativity, the scalar curvature

$$R = g^{\mu\nu} R^\rho{}_{\mu\rho\nu} \tag{17.7}$$

refers now to the curvature of a general Cartan connection:

$$R^\rho{}_{\mu\lambda\nu} = \partial_\lambda \Gamma^\rho{}_{\mu\nu} - \partial_\nu \Gamma^\rho{}_{\mu\lambda} + \Gamma^\rho{}_{\eta\lambda} \Gamma^\eta{}_{\mu\nu} - \Gamma^\rho{}_{\eta\nu} \Gamma^\eta{}_{\mu\lambda}. \tag{17.8}$$

Since connection $\Gamma^\rho{}_{\lambda\mu}$, on account of its non-vanishing torsion, is not symmetric in the last two indices, the Ricci curvature tensor $R_{\mu\nu} = R^\rho{}_{\mu\rho\nu}$ is not symmetric either:

$$R_{\mu\nu} \neq R_{\nu\mu}.$$

Considering the total lagrangian

$$\mathscr{L} = \mathscr{L}_{EC} + \mathscr{L}_s, \tag{17.9}$$

with $\mathscr{L}_s$ the lagrangian of a source field Ψ, the gravitational field equations are obtained by taking variations with respect to the metric $g^{\mu\nu}$ and the contortion tensor $K_\rho{}^{\mu\nu}$. The resulting field equations are

$$R_{\mu\nu} - \tfrac{1}{2} g_{\mu\nu} R = \frac{8\pi G}{c^4} \theta_{\mu\nu} \tag{17.10}$$

and

$$T^\rho{}_{\mu\nu} + \delta^\rho_\mu T^\alpha{}_{\nu\alpha} - \delta^\rho_\nu T^\alpha{}_{\mu\alpha} = \frac{8\pi G}{c^4} s^\rho{}_{\mu\nu}. \tag{17.11}$$

In these equations,

$$\sqrt{-g}\,\theta_\mu{}^\nu = \frac{\partial \mathcal{L}_s}{\partial(\mathcal{D}_\nu \Psi)} h^a{}_\mu \partial_a \Psi - \delta^\nu_\mu \mathcal{L}_s \tag{17.12}$$

is the *canonical* energy-momentum tensor and

$$\sqrt{-g}\, s^\rho{}_{\mu\nu} = \frac{1}{2} \frac{\partial \mathcal{L}_s}{\partial(\mathcal{D}_\rho \Psi)} h^a{}_\mu h^b{}_\nu S_{ab} \Psi \tag{17.13}$$

is the *canonical* spin tensor of the source, with S_{ab} the Lorentz generators taken in the representation to which Ψ belongs. Equation (17.11) can be rewritten in the form

$$T^\rho{}_{\mu\nu} = \frac{8\pi G}{c^4} \left(s^\rho{}_{\mu\nu} + \tfrac{1}{2} \delta^\rho_\mu s^\alpha{}_{\nu\alpha} - \tfrac{1}{2} \delta^\rho_\nu s^\alpha{}_{\mu\alpha} \right). \tag{17.14}$$

We see from this equation that torsion vanishes for spinless ($s^\rho{}_{\mu\nu} = 0$) matter. Concomitantly, the canonical energy-momentum tensor $\theta_\mu{}^\nu$ turns out to be the symmetric energy-momentum tensor $\Theta_\mu{}^\nu$, and the field equation (17.10) reduces to the ordinary Einstein equation. In empty space, therefore, there is no difference between the Einstein-Cartan and Einstein theories. In the presence of spinning matter, however, there will be a non-vanishing torsion, given by Eq. (17.14). As this equation is purely algebraic, torsion is a non-propagating field.

17.3 Gravitational Coupling Prescription

The coupling prescription in Einstein-Cartan theory is defined by

$$\partial_\mu \to \mathcal{D}_\mu = \partial_\mu - \tfrac{i}{2} A^{ab}{}_\mu S_{ab}, \tag{17.15}$$

with S_{ab} an appropriate representation of the Lorentz generators. Acting on a spinor field ψ, for example, it assumes the form

$$\partial_\mu \psi \to \mathcal{D}_\mu \psi = \partial_\mu \psi - \tfrac{i}{2} A^{ab}{}_\mu S_{ab} \psi, \tag{17.16}$$

with S_{ab} the spinor representation

$$S_{ab} = \tfrac{i}{4} [\gamma_a, \gamma_b]. \tag{17.17}$$

Substituting the decomposition (17.3), it becomes

$$\partial_\mu \psi \to \mathcal{D}_\mu \psi = \overset{\circ}{\mathcal{D}}_\mu \psi - \tfrac{i}{2} K^{ab}{}_\mu S_{ab} \psi, \tag{17.18}$$

where

$$\overset{\circ}{\mathscr{D}}_\mu \psi = \partial_\mu \psi - \tfrac{i}{2} \overset{\circ}{A}{}^{ab}{}_\mu S_{ab} \psi \tag{17.19}$$

is the covariant derivative defining the coupling prescription of General Relativity. We see from Eq. (17.18) that new physical phenomena in relation to General Relativity (and consequently with respect to Teleparallel Gravity) are expected in the presence of spin or, equivalently, in the presence of torsion. These additional effects are related to the fact that curvature and torsion represent, in this theory, independent gravitational degrees of freedom.

For a Lorentz scalar field ϕ the generators are null, $S_{ab}\phi = 0$. Furthermore, since torsion vanishes for the (spinless) scalar field, the coupling prescription coincides with that of General Relativity:

$$\partial_\mu \phi \rightarrow \mathscr{D}_\mu \phi = \partial_\mu \phi. \tag{17.20}$$

In the case of a Lorentz vector ϕ^a, on the other hand, for which the generators S_{ab} are given by

$$(S_{ab})^c{}_d = i \left(\eta_{bd} \delta^c_a - \eta_{ad} \delta^c_b \right), \tag{17.21}$$

the coupling prescription (17.16) reads

$$\partial_\mu \phi^a \rightarrow \mathscr{D}_\mu \phi^a = \partial_\mu \phi^a + A^a{}_{b\mu} \phi^b. \tag{17.22}$$

For the corresponding spacetime vector $\phi^\rho = h_a{}^\rho \phi^a$, the coupling prescription has the form

$$\partial_\mu \phi^\rho \rightarrow \nabla_\mu \phi^\rho = \partial_\mu \phi^\rho + \Gamma^\rho{}_{\nu\mu} \phi^\nu, \tag{17.23}$$

or equivalently,

$$\partial_\mu \phi^\rho \rightarrow \nabla_\mu \phi^\rho = \partial_\mu \phi^\rho + \overset{\circ}{\Gamma}{}^\rho{}_{\nu\mu} \phi^\nu + K^\rho{}_{\nu\mu} \phi^\nu, \tag{17.24}$$

where we have used the decomposition (17.5). This is the prescription to be used, for example, in the study of the Einstein-Cartan coupling of the electromagnetic field to gravitation. Observe that this prescription differs from the teleparallel coupling prescription (12.46).

17.4 Particle Equations of Motion

According to the Einstein-Cartan construction, torsion vanishes in absence of spin, and the theory reduces to General Relativity. As a consequence, a spinless particle must satisfy the geodesic equation

$$\frac{du^\rho}{ds} + \overset{\circ}{\Gamma}{}^\rho{}_{\mu\nu} u^\mu u^\nu = 0. \tag{17.25}$$

This equation can be obtained from the free-particle equation of motion

$$u^\nu \partial_\nu u^\rho \equiv \frac{du^\rho}{d\sigma} = 0 \tag{17.26}$$

by applying the coupling prescription (17.24) with $K^{\rho}{}_{\nu\mu} = 0$, as a spinless particle does not produce torsion. Then comes the question: what is the equation of motion of a spinning particle in the context of Einstein-Cartan theory? The most natural solution would be to assume the auto-parallel equation

$$\frac{du^\rho}{ds} + \Gamma^{\rho}{}_{\mu\nu} u^\mu u^\nu = 0, \tag{17.27}$$

which is consistent with the coupling prescription (17.23). However, this equation cannot be the correct because, according to it, all particles are equally affected by torsion, independently of their spin content.

Comment 17.1 There are actually more problems with the equation of motion (17.27). It has already been shown that it cannot be obtained from a lagrangian formalism [6], which means that a spinless particle following such a trajectory does not have a lagrangian. Taking into account that the energy-momentum density is defined as the functional derivative of the lagrangian with respect to the metric tensor (or equivalently, with respect to the tetrad field), it is not possible to define an energy-momentum density for such particle.

One then has to look for another equation of motion for spinning particles. A possible procedure is to begin by observing that the geodesic equation (17.25) is obtained from the lagrangian

$$\mathscr{S} = -\int_a^b h^a{}_\mu p_a \, dx^\mu, \tag{17.28}$$

where $p_a = mcu_a$ is the particle four-momentum. Relying in the Einstein-Cartan coupling prescription (17.22), it is then natural to assume that the action integral describing a spinning particle minimally coupled to a general Lorentz connection $A^a{}_{b\mu}$ be written as

$$\mathscr{S} = \int_a^b \left(-h^a{}_\mu p_a + \tfrac{1}{2} A^{ab}{}_\mu s_{ab}\right) dx^\mu, \tag{17.29}$$

where s_{ab} is the spin angular momentum of the particle. Following the same approach of Appendix A, the routhian describing a spinning particle in Einstein-Cartan theory is written as

$$\mathscr{R} = -h^a{}_\mu p_a u^\mu + \tfrac{1}{2} A^{ab}{}_\mu s_{ab} u^\mu - \frac{\mathscr{D} u^a}{\mathscr{D} s} \frac{s_{ab} u^b}{u^2}, \tag{17.30}$$

where

$$\frac{\mathscr{D} u^a}{\mathscr{D} s} = u^\mu \mathscr{D}_\mu u^a,$$

with $\mathscr{D}_\mu$ the covariant derivative (17.22). The last term of the routhian (17.30) is introduced to ensure that the four-velocity and the spin angular momentum density satisfy the constraints

$$s_{ab} s^{ab} = 2\mathbf{s}^2 \tag{17.31}$$

$$s_{ab} u^a = 0, \tag{17.32}$$

with **s** the particle spin vector. Using this routhian, the equation of motion for the spin is found to be

$$\frac{\mathscr{D}s_{ab}}{\mathscr{D}s} = (u_a s_{bc} - u_b s_{ac})\frac{\mathscr{D}u^c}{\mathscr{D}s}. \tag{17.33}$$

On the other hand, making use of the lagrangian formalism, the equation of motion for the trajectory of the particle is found to be

$$\frac{\mathscr{D}\mathscr{P}_\mu}{\mathscr{D}s} = T^a{}_{\mu\nu}\mathscr{P}_a u^\nu - \tfrac{1}{2}R^{ab}{}_{\mu\nu}s_{ab}u^\nu, \tag{17.34}$$

where $\mathscr{P}_\mu = h_\mu{}^c\mathscr{P}_c$, with

$$\mathscr{P}_c = mcu_c + u^a\frac{\mathscr{D}s_{ca}}{\mathscr{D}s} \tag{17.35}$$

the generalized momentum. This is the Einstein-Cartan version of the Papapetrou equation [7]. In addition to the usual Papapetrou coupling between the particle spin and the Riemann tensor, there is also a coupling between torsion and the generalized momentum $\mathscr{P}_\rho$, which gives rise to new phenomena not predicted by General Relativity—or equivalently, by Teleparallel Gravity. This equation should be compared with the teleparallel and the general relativistic versions of the Papapetrou equation, given respectively by (A.20) and (A.23), in which curvature and torsion appear as alternative ways of describing the same gravitational field.

17.5 Some Drawbacks

The Einstein-Cartan theory, briefly described in this chapter, shows a series of conceptual difficulties. The first one refers to its coupling prescription, defined by Eq. (17.15): it violates the strong equivalence principle (or the general covariance principle), according to which the gravitational coupling prescription is *minimal* only in the spin connection of General Relativity [see Sect. 5.2.1]. Another problem of the same coupling prescription is that, when used to describe the interaction of the electromagnetic field with gravitation, it violates the gauge invariance of the coupled Maxwell equations. In fact, under a gauge transformation of the electromagnetic potential

$$A'_\mu = A_\mu - \partial_\mu\varepsilon, \tag{17.36}$$

the gravitationally-coupled electromagnetic field strength

$$F_{\mu\nu} = \nabla_\mu A_\nu - \nabla_\nu A_\mu, \tag{17.37}$$

with ∇_μ the covariant derivative (17.24), is easily seen not to be gauge invariant:

$$F'_{\mu\nu} = F_{\mu\nu} + T^\rho{}_{\mu\nu}\partial_\rho\varepsilon. \tag{17.38}$$

Furthermore, since the canonical energy-momentum tensor of the electromagnetic field, given by

$$\theta_{\mu\nu} = -\partial_\mu A^\rho F_{\nu\rho} + \tfrac{1}{4} g_{\mu\nu} F_{\rho\sigma} F^{\rho\sigma}, \tag{17.39}$$

is not gauge invariant, when the electromagnetic field is considered as source, the gravitational field equation (17.10) will not be gauge invariant either. Of course, the symmetric energy-momentum tensor

$$\Theta_{\mu\nu} = -F_\mu{}^\rho F_{\nu\rho} + \tfrac{1}{4} g_{\mu\nu} F_{\rho\sigma} F^{\rho\sigma} \tag{17.40}$$

is gauge invariant, but since the left-hand side of the field equation (17.10) is not symmetric in the presence of torsion, such tensor cannot appear as source in the Einstein-Cartan theory. These problems are usually circumvented by *postulating* that the electromagnetic field does not couple nor produce torsion [8–10]. This procedure, however, cannot be considered reasonable in the sense that, instead of giving a solution to the problem, it says how Nature should behave in order not to violate the gauge invariance of the model.

Comment 17.2 Furthermore, from a quantum point of view one may always expect an interaction between photons and torsion [11]. The reason is that a photon, perturbatively speaking, can virtually disintegrate into an electron-positron pair. Considering that these particles are massive fermions that do couple to torsion, a photon will necessarily feel the presence of torsion. Since all macroscopic phenomena must have an interpretation based on an average of microscopic phenomena, and taking into account the strictly attractive character of gravitation, which eliminates the possibility of a vanishing average, the photon field must interact with torsion through the virtual pair produced by the vacuum polarization.

Since the works by Sciama and Kibble in the early nineteen-sixties, in spite of the difficulties, Einstein-Cartan-Sciama-Kibble models have attracted a lot of attention, becoming a subject of intense research. An overview of these works, as well as the relevant literature, can be found in Ref. [12].

References

1. Cartan, É.: Ann. Éc. Norm. Super. **40**, 325 (1923)
2. Cartan, É.: Ann. Éc. Norm. Super. **41**, 1 (1924)
3. Sciama, D.W.: Rev. Mod. Phys. **36**, 463 (1964)
4. Kibble, T.W.B.: J. Math. Phys. **2**, 212 (1961)
5. de Sabbata, V., Gasperini, M.: Introduction to Gravitation. World Scientific, Singapore (1985)
6. Maulbetsch, C., Shabanov, S.V.: J. Phys. A **32**, 5355 (1999)
7. Trautman, A.: Einstein-Cartan theory. In: Françoise, J.P., Naber, G.L., Tsun, T.S. (eds.) Encyclopedia of Mathematical Physics, vol. 2, p. 189. Elsevier, Amsterdam (2006)
8. Benn, I.M., Dereli, T., Tucker, R.W.: Phys. Lett. B **96**, 100 (1980)
9. Hehl, F.W., McCrea, J.D., Mielke, E.W., Ne'eman, Y.: Phys. Rep. **258**, 1 (1995)
10. Shapiro, I.L.: Phys. Rep. **357**, 113 (2002)
11. de Sabbata, V., Sivaram, C.: Spin and Torsion in Gravitation. World Scientific, Singapore (1994)
12. Hehl, F.W.: Gauge theory of gravity and spacetime. Talk given at the workshop Towards a Theory of Spacetime Theories, Bergische Universität Wuppertal, Germany, 21–23 July 2010. 1204.3672 [gr-qc]

Chapter 18
Why to Study Teleparallel Gravity

A general discussion of the distinctive features and achievements of Teleparallel Gravity.

18.1 On the Gravitational Interaction

Gravitation has, at least at the classical level, a quite peculiar property: structureless particles with different masses and compositions experience it in such a way that all of them acquire the same acceleration and, given the same initial conditions, follow the same path. This universality of response—usually referred to as *universality of free fall*—is embodied in the *weak equivalence principle*. It is the most peculiar characteristic of gravity, and unique: no other fundamental interaction of Nature exhibits it. Nevertheless, non-gravitational effects equally felt by all bodies were known since long: they are the *inertial* effects, which show up in non-inertial frames. Examples on Earth are the centrifugal and the Coriolis forces.

Universality of both gravitational and inertial effects was a conceptual clue used by Einstein in building up General Relativity. Another pointer was the notion of field, which provides the best approach to interactions consistent with Special Relativity: all known forces are mediated by fields on spacetime. If also gravitation is to be represented by a field, that field should, by the considerations above, be universal, equally felt by every particle. It is then natural to assume that gravitation changes spacetime itself, and the simplest way to do that is to change what appears as its most fundamental field—the metric. The presence of a gravitational field should be, therefore, represented by a change in the metric of Minkowski spacetime.

The metric tensor, however, defines neither curvature nor torsion by itself. As a matter of fact, curvature and torsion require a *connection* to be defined, and many different connections, with different curvature and torsion tensors, can be defined on the very same metric spacetime. The question then arises: how can we determine the relevant connection to describe the gravitational field? This is a fundamental question, which has more than one answer. For example, when constructing General Relativity, Einstein chose the torsionless Levi-Civita, or Christoffel connection,

R. Aldrovandi, J.G. Pereira, *Teleparallel Gravity*, Fundamental Theories of Physics 173, 179
DOI 10.1007/978-94-007-5143-9_18,

which is a connection completely specified by the ten components of the metric tensor. Its curvature represents the gravitational field. In this theory, therefore, torsion is chosen to vanish from the very beginning.

This is not, however, the only possible choice. A second possibility is to choose the Weitzenböck connection, a zero-curvature Lorentz connection not related to gravitation, but to inertial effects only. The gravitational theory that emerges from this choice is Teleparallel Gravity, a gauge theory for the translation group, in which curvature is assumed to vanish from the very beginning. In this theory, the gravitational field turns out to be represented by a translational gauge potential, which appears as the non-trivial part of the tetrad field and gives rise to a non-vanishing torsion. As a gauge theory, the gravitational interaction is described by a force, and the particle trajectories are not geodesics, but force equations with torsion (or contortion) playing the role of force.

Einstein's choice, it must be said, is the most intuitive from the point of view of universality. Gravitation can be easily understood by supposing that it produces a curvature in spacetime, in such a way that all (spinless, structureless) particles, independently of their masses and constitutions, will follow a geodesic on the curved spacetime. Universality of free fall is, in this way, naturally incorporated into gravitation. Geometry replaces the concept of force and the trajectories are solutions, not of a force equation, but of a geodesic equation. Nevertheless, because such a geometrization relies on the weak equivalence principle, in absence of universality the general-relativistic description of gravitation would break down.

This restriction apart, one may wonder whether there is any problem with Einstein's choice. Was Einstein wrong when he chose a torsionless connection to describe gravitation? This question is as old as General Relativity. In fact, since the early days of this theory, there have been theoretical speculations on the necessity of including torsion, *in addition to curvature*, in the description of the gravitational interaction. This comes from the fact that a Lorentz connection, as described in its generality by Cartan, presents naturally *both* curvature and torsion. Theories like Einstein-Cartan [1, 2] and gauge models for the Poincaré group [3, 4] consider curvature and torsion as representing independent degrees of freedom. New effects should become relevant at the microscopic level, when spins are important. And new physical phenomena are expected in the presence of torsion. From the point of view of these theories, Einstein would have made a mistake by neglecting torsion.

On the other hand, although conceptually different, General Relativity and Teleparallel Gravity are found to yield equivalent descriptions of the gravitational interaction. An immediate implication of this equivalence is that curvature and torsion turn out to be just alternative ways of describing the gravitational field. This is corroborated by the fact that the symmetric matter energy-momentum tensor appears as the source in both theories: of curvature in General Relativity, of torsion in Teleparallel Gravity. According to this interpretation, both General Relativity and Teleparallel Gravity are complete theories, and Einstein did not make any mistake by leaving torsion aside.

18.2 On Teleparallel Gravity

Teleparallel Gravity is an alternative theory, fully equivalent to General Relativity. In spite of this equivalence, it shows many distinguished features that make of it a theory worthy studying. Some of these features are reviewed and discussed below.

18.2.1 A Gauge Theory for Gravitation

Although equivalent to General Relativity, Teleparallel Gravity gives of gravitation a completely different picture. Curvature is replaced by torsion, geometry by force. Behind this difference lies the gauge structure: teleparallelism shows up as a gauge theory for the group of spacetime translations, which explains why gravitation has for source energy-momentum, just the Noether current for those translations. Soldering makes of it a non-standard gauge theory, keeping nevertheless a remarkable similarity to electromagnetism, also a gauge theory for an abelian group. Considering that the other known interactions are described by gauge theories, Teleparallel Gravity brings gravitation to the fold. This is just a formal advantage, but important for the understanding of Nature. Gravitation is no longer a different interaction; only General Relativity is a different theory.

Comment 18.1 Starting with the pioneering work of Utiyama in 1956 [5], there have been many attempts to construct a gauge theory for gravitation. Most of the models are extensions of General Relativity, in which curvature and torsion (and sometimes even non-metricity) are related to different degrees of freedom, and consequently to different physical phenomena. General Relativity is always recovered as a particular case. A sample of references covering these attempts, from where the relevant literature ca be traced back, is displayed in Refs. [6–10].

18.2.2 Matters of Consistency

As of today, there are no experimental evidences for *new physics* associated to torsion. This is true at both microscopic and macroscopic levels. At the microscopic level, present-day experimental sensitivity is well below the necessary to detect any effect related to the spin-torsion coupling predicted by Einstein-Cartan theory [11]. At the macroscopic level, an interesting laboratory would be a neutron star, where the alignment of the neutrons should produce a macroscopic spin, and consequently a torsion field. However, the existence and stability of neutron stars are well understood on the basis of General Relativity. We can then say that the teleparallel point of view is favored by the available experimental data. Furthermore, all known gravitational phenomena, including the physics of the solar system, can be consistently re-interpreted in terms of the teleparallel force equation, with contortion as force. Though unnoticed by many, therefore, teleparallel torsion has already been detected.

In addition to the phenomenological consistency, Teleparallel Gravity shows also a sound conceptual consistency. For example, the gravitational coupling prescription of models like Einstein-Cartan violates the strong equivalence principle, and when used to describe the gravitational interaction of the electromagnetic field, violates the $U(1)$ gauge invariance of Maxwell theory. On the other hand, the coupling prescription of Teleparallel Gravity, like that of General Relativity, is consistent with both the active and passive versions of the strong equivalence principle, and when applied to describe the gravitational interaction of the electromagnetic field, it is found not to violate the $U(1)$ gauge invariance of Electromagnetism. As this invariance is of paramount importance to physics, the torsion interpretation provided by Teleparallel Gravity can be considered as the most natural in the sense that it is consistent with well-established theories.

18.2.3 Gravitational Energy-Momentum Density

All fundamental fields have well-defined local energy-momentum densities. It would then be natural to expect that the same should happen to the gravitational field. It is true, however, that no tensorial expression for the gravitational energy-momentum density can be defined in the context of General Relativity. The basic reason for this impossibility is that gravitational and inertial effects are mixed in the spin connection of the theory, and cannot be separated. Even though some quantities, like for example curvature, are not affected by inertia, some others turn out to depend on it. For example, the energy-momentum density of gravitation will necessarily include both a contribution from gravity and another from the inertial effects. Since the inertial effects are non-tensorial by its very nature, it is not surprising that in this theory any complex defining the energy-momentum density of the gravitational field always shows up as a non-tensorial object.

On the other hand, while inertial effects in Teleparallel Gravity remain described by a Lorentz connection, gravitation turns out to be represented by a translational gauge potential, which appears as the non-trivial part of the tetrad. This theory thus naturally separates gravitation from inertial effects. As a consequence, it becomes possible to write down an energy-momentum density for gravitation only, excluding the contribution from inertia. This object is a true tensor, which means that *gravitation, like any other field of nature, does have a tensorial energy-momentum definition*. Since it does not represent the total energy-momentum density—in the sense that the inertial-related part is not included—it is not truly conserved, but only covariantly conserved. Of course, the total energy-momentum density, which in the general case includes contributions from inertia, gravitation and matter, remains conserved in the ordinary sense. We can then say that the impossibility of defining a tensorial expression for the gravitational energy-momentum density is not a characteristic of gravity, but a property of the geometrical picture of General Relativity.

18.2.4 Coupling of a Fundamental Spin-2 Field to Gravitation

It is well known that a fundamental spin-2 field presents inconsistencies when coupled to gravitation. The problem is that the divergence identities satisfied by the field equations of a spin-2 field in Minkowski spacetime are no longer valid when it is coupled to gravitation. The basic underlying difficulty is related to the fact that the covariant derivative of General Relativity—which defines the gravitational coupling prescription—is non-commutative, and this introduces unphysical constraints on the spacetime curvature. Furthermore, since the coupled equations are no longer gauge invariant, the spurious components of the field cannot be eliminated.

The dynamics of a spin-2 field in Minkowski spacetime is expected to coincide with the dynamics of a linear perturbation of the metric around flat spacetime. For this reason, a fundamental spin-2 field is usually assumed to be described by a rank-two, symmetric tensor $\psi_{\mu\nu} = \psi_{\nu\mu}$. However, conceptually speaking, this is not the most fundamental notion of a spin-2 field. In fact, remember that the description of a gravitationally-coupled spinor field requires the use of the tetrad formalism [12]. The tetrad formalism, therefore, can be considered to be more fundamental than the metric formulation in the sense that it is able to describe the gravitational interaction of both tensor and spinor fields. Accordingly, instead of similar to a linear perturbation of the metric, a fundamental spin-2 field should be interpreted as a linear perturbation $\phi^a{}_\mu$ of the tetrad, and consequently as a 1-form assuming values in the Lie algebra of the translation group.

When the spin-2 field is interpreted as a translational-valued 1-form, and the teleparallel paradigm is used to describe its dynamics, a sound spin-2 field theory emerges, which is quite similar to the gravitationally-coupled electromagnetic theory. It is both gauge and local-Lorentz invariant, and it preserves the duality symmetry of the free theory. In addition, since the teleparallel spin connection is purely inertial, no unphysical constraints on the spacetime geometry shows up. This property, together with the gauge and local Lorentz invariance, render the teleparallel-based gravitationally-coupled spin-2 theory fully consistent.

18.2.5 Gravitation and Quantum Mechanics

Owing to its gauge structure, Teleparallel Gravity dispenses with the weak equivalence principle. It can comply with universality, but remains a consistent theory in its absence. This property can have deeper consequences. For example, it is well known that General Relativity and Quantum Mechanics are not consistent with each other. This conflict stems from the very principles on which these theories take their roots. General Relativity is based on the equivalence principle, whose strong version establishes the *local* equivalence between gravitation and inertia. The fundamental asset of Quantum Mechanics, on the other hand, is the uncertainty principle, which is essentially *nonlocal*: a test particle does not follow a given trajectory, but infinitely many trajectories, each one with a different probability.

The question then arises: is there a consistent way of reconciling the equivalence and the uncertainty principles? To begin with, observe that the strong equivalence principle presupposes an ideal observer [13], represented by a timelike curve which intersects the space-section *at a point*. In each space-section, it applies at that intersecting point. The conflict comes from that idealization and extends, clearly, also to Special Relativity. In the geodesic equation, gravitation only appears through the Levi-Civita connection, which can be made to vanish all along the curve. An *ideal* observer can choose frames whose acceleration exactly compensate the effect of gravitation. A *real* observer, on the other hand, will be necessarily an object extended in space, consequently intersecting a congruence of curves. Such congruences are described by the geodesic deviation equation and, consequently, detect the true covariant object characterizing the gravitational field, the curvature tensor—which cannot be made to vanish. Quantum Mechanics requires real observers, pencils of ideal observers. The inconsistency with the strong principle, therefore, is a mathematical necessity which precludes the existence of a quantum version of the strong equivalence principle.

It seems, therefore, that the equivalence and the uncertainty principles cannot hold simultaneously. Then comes the point: although the geometrical description of General Relativity depends on the equivalence principle, the gauge structure of Teleparallel Gravity does not require it to describe gravitation. As a consequence, provided Teleparallel Gravity is used, one can dispense with the equivalence principle. Synge's old injunction [14] to the effect that *the midwife be now buried with appropriate honours* can thus be carried out. Of course, this does not mean that the inconsistencies between gravitation and Quantum Mechanics are automatically solved. It means simply that Teleparallel Gravity provides a better framework than General Relativity to deal with this problem.

18.2.6 Quantizing Gravity

The spin connection of General Relativity represents both gravitation and inertial effects. For this reason, it is not a genuine gravitational variable in the usual sense of field theory. It is not a gravitational connection either in the sense that its connection behavior is due to its inertial content, not to gravitation itself—its gravitational content is covariant. One should not expect, therefore, any dynamical effect coming from a "gaugefication" of the Lorentz group. As a matter of fact, local Lorentz transformations are related to different classes of non-inertial frames. Due to the local equivalence between the inertial effects of these frames and gravitation, they are actually related to the geometric description of General Relativity, not to any gauge approach.

In Teleparallel Gravity, on the other hand, while Lorentz connections keep their special-relativistic role of describing inertial effects only, gravitation is represented by a translational-valued gauge potential, which shows up as the non-trivial part of the tetrad field. Since it does not include inertial effects, it is a true gravitational connection, and consequently a genuine field variable in the usual sense of field theory.

This is clear from the fact that, differently from the spin connection of General Relativity, it cannot be made to vanish in a point through the choice of an appropriate local frame. It is, for this reason, the natural field-variable to be quantized in any approach to quantum gravity.

18.2.7 A New Insight into Gravity

Owing to its uncompensated long range, gravitation is the interaction presiding over the large scale structure and evolution of the Universe. Any change in the way we interpret the gravitational interaction will produce concomitant changes in the very way we see and conceive the Cosmos. For example, according to the geometric description of General Relativity, which makes use of the torsionless Levi-Civita connection, there is a widespread belief that gravity produces a curvature in spacetime. In consequence, the Universe as a whole should be curved. However, the advent of Teleparallel Gravity changes this concept. In fact, because of the equivalence between Teleparallel Gravity and General Relativity, it becomes a matter of convention to describe the gravitational interaction in terms of curvature or in terms of torsion. This means that the attribution of curvature to spacetime is not an absolute, but a model-dependent statement. Of course, Cosmology as based on General Relativity is not at all incorrect. However, an appraisal based on Teleparallel Gravity could provide a new frame of mind to observe and interpret the Cosmos.

Not only Cosmology, but all gravitational phenomena would acquire a new perspective when analyzed from the teleparallel point of view. For instance, seen from the teleparallel point of view, the age-old problem of the energy-momentum density of the gravitational field has found a solution. Also the problem of the gravitational coupling of a fundamental spin-2 field has found a reasonable solution. In the same token, gravitational waves would no longer be interpreted as curvature perturbations in the fabric of spacetime, but as field-strength waves, like in electromagnetism. Similarly, other gravitational phenomena could find new interpretations in the teleparallel paradigm. We can then say that Teleparallel Gravity is not just a theory which is equivalent to General Relativity, but a new way to look at all gravitational phenomena, including those shaping the Universe itself.

References

1. Cartan, É.: Ann. Éc. Norm. Super. **40**, 325 (1923)
2. Cartan, É.: Ann. Éc. Norm. Super. **41**, 1 (1924)
3. Sciama, D.W.: Rev. Mod. Phys. **36**, 463 (1964)
4. Kibble, T.W.B.: J. Math. Phys. **2**, 212 (1961)
5. Utiyama, R.: Phys. Rev. **101**, 1597 (1956)
6. Ivanenko, D., Sardanashvily, G.: Phys. Rep. **94**, 1 (1983)
7. Sardanashvily, G., Zakharov, O.: Gauge Gravitation Theory. World Scientific, Singapore (1992)

8. Hehl, F.W., McCrea, J.D., Mielke, E., Ne'eman, Y.: Phys. Rep. **258**, 1 (1995)
9. Blagojević, M.: Gravitation and Gauge Symmetries. IOP Publishing, Bristol (2002)
10. Blagojević, M., Hehl, F.W. (eds.): Gauge Theories of Gravitation, a reader with commentaries. Imperial College Press, London (2012).
11. Trautman, A.: Einstein-Cartan theory. In: Françoise, J.P., Naber, G.L., Tsun, T.S. (eds.) Encyclopedia of Mathematical Physics, vol. 2, p. 189. Elsevier, Amsterdam (2006)
12. Veltman, M.J.G.: Quantum theory of gravitation. In: Balian, R., Zinn-Justin, J. (eds.) Methods in Field Theory, Les Houches 1975. North-Holland, Amsterdam (1976)
13. Aldrovandi, R., Barros, P.B., Pereira, J.G.: Found. Phys. **33**, 545 (2003). gr-qc/0212034
14. See the preface of Synge, J.L.: Relativity: The General Theory. Wiley, New York (1960)

Appendix A
The Spinning Particle

The classical equation of motion of a spinning particle of mass m and spin **s** in the presence of gravitation is obtained in the context of Teleparallel Gravity. Its equivalence with the Papapetrou equation is established.

A.1 Action Functional

In Chap. 6 the equation of motion of a spinless particle of mass m in the presence of gravitation was obtained. We consider now the motion of a particle of mass m and spin **s** in the presence of gravitation. In the context of Teleparallel Gravity, the action integral describing such a particle minimally coupled to gravitation is

$$\mathscr{S} = \int_{\mathsf{p}}^{\mathsf{q}} \big[-(\partial_\mu x^a + \overset{\bullet}{A}{}^a{}_{b\mu} x^b + B^a{}_\mu) p_a + \tfrac{1}{2} (\overset{\bullet}{A}{}^{ab}{}_\mu - \overset{\bullet}{K}{}^{ab}{}_\mu) s_{ab} \big] dx^\mu, \tag{A.1}$$

where $p_a = mcu_a$ is the Noether charge associated with the invariance of $\mathscr{S}$ under spacetime translations, and $s_{ab} = -s_{ba}$ is the Noether charge associated with the invariance of $\mathscr{S}$ under Lorentz transformations [1, 2]. In other words, p_a is the momentum, and s_{ab} is the spin angular momentum density, which satisfies the Poisson relation

$$\{s_{ab}, s_{cd}\} = \eta_{ac} s_{bd} + \eta_{bd} s_{ac} - \eta_{ad} s_{bc} - \eta_{bc} s_{ad}. \tag{A.2}$$

Notice that, according to the above prescription, the particle momentum couples minimally to the translational gauge potential $B^a{}_\mu$, whereas the spin of the particle, as implied by the general covariance principle [see Sect. 5.2.1], couples minimally to the dynamical spin connection

$$\overset{\bullet}{A}{}^{ab}{}_\mu - \overset{\bullet}{K}{}^{ab}{}_\mu \equiv \overset{\circ}{A}{}^{ab}{}_\mu. \tag{A.3}$$

R. Aldrovandi, J.G. Pereira, *Teleparallel Gravity*, Fundamental Theories of Physics 173,
DOI 10.1007/978-94-007-5143-9, © Springer Science+Business Media Dordrecht 2013

A.2 Equations of Motion

A quite convenient way to get the equations of motion is by using the routhian formalism, according to which the equation of motion for the particle trajectory comes from the Lagrange formulation, and the spin equation of motion comes from the Hamilton formulation. The routhian arising from action (A.1) is

$$\mathscr{R}_0 = -(\partial_\mu x^a + \overset{\bullet}{A}{}^a{}_{b\mu} x^b + B^a{}_\mu) p_a u^\mu + \tfrac{1}{2}(\overset{\bullet}{A}{}^{ab}{}_\mu - \overset{\bullet}{K}{}^{ab}{}_\mu) s_{ab} u^\mu. \tag{A.4}$$

The equation of motion for the particle trajectory is obtained from

$$\frac{\delta}{\delta x^\mu} \int_{\mathsf{p}}^{\mathsf{q}} \mathscr{R}_0 \, ds = 0, \tag{A.5}$$

whereas the equation of motion for the spin tensor follows from

$$\frac{ds_{ab}}{ds} = \{\mathscr{R}_0, s_{ab}\}. \tag{A.6}$$

The four-velocity and the spin angular momentum density must satisfy the constraints

$$s_{ab} s^{ab} = 2\mathbf{s}^2 \tag{A.7}$$

$$s_{ab} u^a = 0, \tag{A.8}$$

with $\mathbf{s}$ the particle spin vector. However, since the equations of motion that are obtained from the routhian $\mathscr{R}_0$ do not satisfy the above constraints, it is necessary to include those constraints in the routhian. The simplest way to achieve this amounts to the following [3]. First, a new expression for the spin is introduced:

$$\tilde{s}_{ab} = s_{ab} - \frac{s_{ac} u^c u_b}{u^2} - \frac{s_{cb} u^c u_a}{u^2}. \tag{A.9}$$

This new tensor satisfies the Poisson relation (A.2) with the metric

$$\eta_{ab} - u_a u_b / u^2. \tag{A.10}$$

A new routhian that incorporates the above constraints is obtained by replacing all s_{ab} in $\mathscr{R}_0$ by $\tilde{s}_{ab}$, and by subtracting from it the term

$$\frac{du^a}{ds} \frac{s_{ab} u^b}{u^2}. \tag{A.11}$$

The result is

$$\mathscr{R} = \mathscr{R}_0 - \frac{\overset{\bullet\bullet}{\mathscr{D}} u^a}{\mathscr{D} s} \frac{s_{ab} u^b}{u^2}, \tag{A.12}$$

where

$$\frac{\overset{\bullet\bullet}{\mathscr{D}} u^a}{\mathscr{D} s} = u^\mu \overset{\bullet\bullet}{\mathscr{D}}_\mu u^a, \tag{A.13}$$

with $\overset{\bullet\bullet}{\mathscr{D}}_\mu$ the covariant derivative (5.34). Using this routhian, the equation of motion for the spin is found to be

$$\frac{\overset{\bullet\bullet}{\mathscr{D}} s_{ab}}{\mathscr{D} s} = (u_a s_{bc} - u_b s_{ac}) \frac{\overset{\bullet\bullet}{\mathscr{D}} u^c}{\mathscr{D} s}, \tag{A.14}$$

which, on account of the equivalence (A.3), coincides with the corresponding result of General Relativity.

Making use of the lagrangian formalism, the next step is to obtain the equation of motion for the trajectory of the particle. Through a tedious but straightforward calculation, one finds

$$\frac{\overset{\bullet\bullet}{\mathscr{D}}}{\mathscr{D} s}(mcu_c) + \frac{\overset{\bullet\bullet}{\mathscr{D}}}{\mathscr{D} s}\left(\frac{\mathscr{D} u^a}{\mathscr{D} s}\frac{s_{ac}}{u^2}\right) = -\tfrac{1}{2}\dot{Q}^{ab}{}_{\mu\nu} s_{ab} u^\nu h_c{}^\mu, \tag{A.15}$$

where [see Eq. (9.26)]

$$\dot{Q}^a{}_{b\mu\nu} = \dot{\mathscr{D}}_\mu \dot{K}^a{}_{b\nu} - \dot{\mathscr{D}}_\nu \dot{K}^a{}_{b\mu} + \dot{K}^a{}_{d\mu}\dot{K}^d{}_{b\nu} - \dot{K}^a{}_{d\nu}\dot{K}^d{}_{b\mu} \tag{A.16}$$

is a curvature-like tensor, which depends on torsion only. Using the constraints (A.7)–(A.8), it is easy to verify that

$$\frac{\overset{\bullet\bullet}{\mathscr{D}} u^a}{\mathscr{D} s}\frac{s_{ac}}{u^2} = u^a \frac{\overset{\bullet\bullet}{\mathscr{D}} s_{ca}}{\mathscr{D} s}. \tag{A.17}$$

As a consequence, Eq. (A.15) acquires the form

$$\frac{\overset{\bullet\bullet}{\mathscr{D}}}{\mathscr{D} s}\left(mcu_c + u^a \frac{\overset{\bullet\bullet}{\mathscr{D}} s_{ca}}{\mathscr{D} s}\right) = -\tfrac{1}{2}\dot{Q}^{ab}{}_{\mu\nu} s_{ab} u^\nu h_c{}^\mu. \tag{A.18}$$

Defining the generalized four-momentum

$$\mathscr{P}_c = h_c{}^\mu \mathscr{P}_\mu \equiv mcu_c + u^a \frac{\overset{\bullet\bullet}{\mathscr{D}} s_{ca}}{\mathscr{D} s}, \tag{A.19}$$

we get

$$\frac{\overset{\bullet\bullet}{\mathscr{D}} \mathscr{P}_\mu}{\mathscr{D} s} = -\tfrac{1}{2}\dot{Q}^{ab}{}_{\mu\nu} s_{ab} u^\nu. \tag{A.20}$$

This is the teleparallel version of the Papapetrou equation [4]. Notice that the particle spin, similarly to the electromagnetic field [see the teleparallel Maxwell equation (12.55)], couples to a curvature-like tensor, which is however a tensor written in terms of torsion only.

When the spin vanishes, the equation of motion (A.20) reduces to

$$\frac{\overset{\bullet\bullet}{\mathscr{D}}}{\mathscr{D} s}(mcu_c) = 0, \tag{A.21}$$

which is just the teleparallel force equation for spinless particles

$$\frac{du_c}{ds} - \dot{A}^b{}_{c\rho} u_b u^\rho = \dot{K}^b{}_{c\rho} u_b u^\rho. \tag{A.22}$$

On the other hand, when rewritten in terms of the spin connection of General Relativity, the teleparallel equation of motion (A.20) reduces to the ordinary Papapetrou equation [5]

$$\frac{\overset{\circ}{\mathscr{D}}\mathscr{P}_\mu}{\mathscr{D}s} = -\frac{1}{2}\overset{\circ}{R}{}^{ab}{}_{\mu\nu} s_{ab} u^\nu. \tag{A.23}$$

References

1. Wong, S.K.: Nuovo Cimento A **65**, 689 (1970)
2. Drechsler, W.: Phys. Lett. B **90**, 258 (1980)
3. Yee, K., Bander, M.: Phys. Rev. D **48**, 2797 (1993)
4. Arcos, H.I., de Andrade, V.C., Pereira, J.G.: Int. J. Mod. Phys. D **13**, 807 (2004). gr-qc/0403074
5. Papapetrou, A.: Proc. R. Soc. Lond. Ser. A **209**, 248 (1951)

Appendix B
The Connection Space

The space of Lorentz connections is an affine space. Different points of this space define different connections, not related by Lorentz transformations. Each connection defines an acceleration, and provides a particular way of describing the gravitational interaction. Whereas the spin connection of General Relativity defines a vanishing acceleration, and then the geodesic equation, the spin connection of Teleparallel Gravity defines an acceleration which is related to a purely gravitational force. In between there are infinitely many connections, each one describing the gravitational interaction partially as geometry, partially as force.

B.1 The Space of Lorentz Connections

The space of connections is an infinite, homotopically trivial affine space [1]. This means that each point on any straight line drawn through two connections is also a connection. Given two connections $A_{(0)}$ and $A_{(1)}$,

$$A_{(\xi)} = \xi A_{(1)} + (1 - \xi) A_{(0)} \tag{B.1}$$

will be a connection for any real value of ξ. Connections are not covariant objects, but the difference between two connections is a tensor, an object that changes covariantly under transformations. These properties hold for connections defined on both internal and external bundles. Gauge connections of the Yang-Mills type belong to the first set, whereas Lorentz and translational connections belong to the second set.

Comment B.1 The space of connections is similar to the space of points, which is also an affine space. The difference between two points is not a point, but a vector. Points themselves are not covariant in any sense, but vectors are.

Our interest here will be concentrated on Lorentz connections, which are instrumental in describing the gravitational interaction. A point in the space of Lorentz connections is specified by a connection 1-form

$$A = \tfrac{1}{2} A^{ab}{}_{\mu} S_{ab} \, dx^{\mu}, \tag{B.2}$$

R. Aldrovandi, J.G. Pereira, *Teleparallel Gravity*, Fundamental Theories of Physics 173, DOI 10.1007/978-94-007-5143-9, © Springer Science+Business Media Dordrecht 2013

with S_{ab} the generators in a representation of the Lorentz group. Under a local Lorentz transformation, A transforms non-covariantly,

$$A'_\mu = \Lambda A_\mu \Lambda^{-1} + \Lambda \partial_\mu \Lambda^{-1} \tag{B.3}$$

with Λ an element of the Lorentz group in the vector representation. Notice that, if a connection is zero in one frame, it will not be zero in another frame.

In the general case, a Lorentz connection A has non-vanishing curvature and torsion. In the language of differential forms, these quantities are defined respectively by

$$R = dA + AA \equiv \mathscr{D}_A A \tag{B.4}$$

and

$$T = dh + Ah \equiv \mathscr{D}_A h, \tag{B.5}$$

where $\mathscr{D}_A$ denotes a covariant differential with the connection A, and

$$h = h^a{}_\mu P_a \, dx^\mu \tag{B.6}$$

is a tetrad field. It is interesting to observe that property (B.1) does not transfer to curvature, but transfers to torsion:

$$T_{(\xi)} = \xi T_{(1)} + (1 - \xi) T_{(0)}. \tag{B.7}$$

It transfers to contortion as well,

$$K_{(\xi)} = \xi K_{(1)} + (1 - \xi) K_{(0)}, \tag{B.8}$$

also a 1-form assuming values in the Lorentz Lie algebra,

$$K = \tfrac{1}{2} K^{ab}{}_\mu S_{ab} \, dx^\mu, \tag{B.9}$$

but transforming covariantly under local Lorentz transformations:

$$K'_\mu = \Lambda K_\mu \Lambda^{-1}. \tag{B.10}$$

Its components, we recall, are

$$K^{ab}{}_\mu = \tfrac{1}{2} \left(T^{ba}{}_\mu + T_\mu{}^{ab} - T^{ab}{}_\mu \right), \tag{B.11}$$

with $T^{ab}{}_\mu$ the torsion tensor.

Comment B.2 It is interesting to observe how a combination of the translational-valued 2-form torsion transmutes into the Lorentz-valued 1-form contortion.

Given two Lorentz connections A and $\bar{A}$, the difference

$$\bar{A} - A = K \tag{B.12}$$

is not a connection, but a contortion 1-form. Using the definition (B.4) for both A and $\bar{A}$, one can verify that the relation between the respective curvatures is

$$\bar{R} = R + \mathscr{D}_A K + KK \tag{B.13}$$

where

$$\mathscr{D}_A K = dK + \{A, K\} \tag{B.14}$$

is the contortion covariant derivative. On the other hand, using the definition (B.5), one can see that

$$\bar{T} = T + Kh. \tag{B.15}$$

The effect of adding a covector K to a given connection A is, then, to change its curvature and torsion 2-forms. The space of Lorentz connections can then be said to be transitive under "contortion translations"

$$A \to \bar{A} = A + K. \tag{B.16}$$

Comment B.3 The fact that connections belong to affine spaces allows a detailed examination of an interesting question: the Wu-Yang ambiguity [2], or the *problem of copies* [3]. The fundamental field of a gauge theory is the potential $A^C{}_\nu$, but the observable, measurable field is the field strength $F^C{}_{\mu\nu}$. It may come as a surprise that, in non-abelian gauge theories, many quite non-equivalent potentials $A^C{}_\nu$ (called copies) can have one same field strength $F^C{}_{\mu\nu}$. In what concerns torsion, there are no copies for lorentzian connections. Let us state in more detail the Ricci theorem [4] mentioned in Comment 1.9: given a metric $g_{\mu\nu}$ and any tensor of type $T^\rho{}_{\mu\nu}$, there exists one and only one linear connection Γ which preserves $g_{\mu\nu}$ and has torsion equal to $T^\rho{}_{\mu\nu}$. In particular, the only lorentzian connection with $T^\rho{}_{\mu\nu} = 0$ is the Levi-Civita connection, from which the others differ precisely by their torsions. Lorentzian connections are, in this way, *classified* by their torsions. General Relativity, by postulating $T^\rho{}_{\mu\nu} = 0$, chooses the Levi-Civita connection once and for all, thereby avoiding the copies problem. Notice that this is still another consequence of the existence of torsion, even if vanishing.

B.2 Equivalence Under Contortion Translations

Let us choose the point on the space of connections representing the vanishing connection

$$\overset{\bullet}{A}{}^a{}_{bc} = 0, \tag{B.17}$$

which is just the spin connection of Teleparallel Gravity written in the special class of inertia-free frames. Seen from a general Lorentz-rotated frame, this connection has the form

$$\overset{\bullet}{A}{}^a{}_{bc} = \Lambda^a{}_e h_c \Lambda_b{}^e. \tag{B.18}$$

Performing a translation with parameter $-\overset{\bullet}{K}{}^a{}_{bc}$, we obtain the new connection

$$\overset{\bullet}{A}{}^a{}_{bc} \to \overset{\bullet}{A}{}^a{}_{bc} - \overset{\bullet}{K}{}^a{}_{bc} = \overset{\circ}{A}{}^a{}_{bc}, \tag{B.19}$$

which is just the spin connection of General Relativity. We see in this way that Teleparallel Gravity and General Relativity are related by a translation in the space of connections. This is actually not a property of the spin connection of Teleparallel Gravity, but an universal property: given a general connection $A^a{}_{bc}$, if one performs a translation using the connection contortion $-K^a{}_{bc}$, one ends up with the spin connection of General Relativity:

$$A^a{}_{bc} - K^a{}_{bc} = \overset{\circ}{A}{}^a{}_{bc}. \tag{B.20}$$

Now, as we have discussed in Sect. 2.3, each connection defines an acceleration defined by Eq. (2.33)—which, by the way, implies that property (B.1) transfers also to accelerations. There are, thus, many different accelerations, one for each point of the connection space. Of course, these connections are in general quite independent, in the sense that they are not related by any Lorentz transformations. This holds, in particular, for connections $\overset{\bullet}{A}{}^a{}_{bc}$ and $\overset{\circ}{A}{}^a{}_{bc}$. In General Relativity, for example, the connection is such that all accelerations vanish identically. The geodesic equation

$$\frac{du^c}{ds} + \overset{\circ}{A}{}^c{}_{ab} u^a u^b = 0 \tag{B.21}$$

is just the mathematical expression of this statement. As a consequence, there is no concept of gravitational force in General Relativity. In Teleparallel Gravity, on the other hand, the spin connection represents inertial effects only, and the resulting acceleration is entirely related to a gravitational force. In fact, the equation of motion is, in this case, the force equation

$$\frac{du^c}{ds} + \overset{\bullet}{A}{}^c{}_{ab} u^a u^b = \overset{\bullet}{K}{}^c{}_{ab} u^a u^b. \tag{B.22}$$

Of course, due to the relation (B.20), the acceleration defined by any connection will give rise to an equivalent equation of motion

$$\frac{du^c}{ds} + A^c{}_{ab} u^a u^b = K^c{}_{ab} u^a u^b. \tag{B.23}$$

In this case, the gravitational interaction is partially described by geometry, and partially described by a force.

Comment B.4 The General Relativity spin connection $\overset{\circ}{A}{}^a{}_{bc}$ is fully determined by the metric tensor, or the tetrad. A general spin connection $A^a{}_{bc}$ with curvature and torsion, on the other hand, has more independent components, and cannot be fully determined by the gravitational field equations. However, since the difference $A^a{}_{bc} - K^a{}_{bc}$ is equivalent to $\overset{\circ}{A}{}^a{}_{bc}$, such difference has the same number of independent components of $\overset{\circ}{A}{}^a{}_{bc}$, and can consequently be determined by the gravitational field equations.

If one wants to quantify how much of the interaction is geometry and how much is gravitational force, one can use expressions (B.1) for the specific case in which

$$A_{(0)} = \overset{\circ}{A} \quad \text{and} \quad A_{(1)} = \overset{\bullet}{A}. \tag{B.24}$$

Namely,

$$A_{(\xi)} = \xi \overset{\bullet}{A}_{(1)} + (1-\xi)\overset{\circ}{A}_{(0)}. \tag{B.25}$$

As a consequence, we have also

$$K_{(\xi)} = \xi \overset{\bullet}{K}_{(1)} + (1-\xi)\overset{\circ}{K}_{(0)}. \tag{B.26}$$

In this case, the equation of motion (B.23) assumes the form

$$\frac{du^c}{ds} + A^c_{(\xi)ab} u^a u^b = K^c_{(\xi)ab} u^a u^b. \tag{B.27}$$

For $\xi = 0$ it reduces to the geodesic equation (B.21), as $\overset{\circ}{K}_{(0)}$ is actually zero; for $\xi = 1$ it reduces to the force equation (B.22). For $0 < \xi < 1$, the gravitational interaction in the equation of motion is partially described by geometry, partially by force.

References

1. Singer, I.M.: Commun. Math. Phys. **60**, 7 (1978)
2. Wu, T.T., Yang, C.N.: Phys. Rev. D **12**, 3845 (1975)
3. Aldrovandi, R., Barbosa, A.L.: Int. J. Math. Math. Sci. **15**, 2365 (2005). math-ph/0410034
4. Halperin, S., Greub, W.H., Vanstone, R.: Connections, Curvature and Cohomology: De Rham Cohomology of Manifolds and Vector Bundles. Academic Press, New York (1972)

Appendix C
Teleparallel Field Equation

The computations leading to the teleparallel field equation of Sect. 9.4 are presented.

C.1 Lagrangian and Field Equation

According to Eq. (9.16), the lagrangian of Teleparallel Gravity is

$$\overset{\bullet}{\mathscr{L}} = \frac{h}{2k}\left(\tfrac{1}{4}\overset{\bullet}{T}{}^{\rho}{}_{\mu\nu}\overset{\bullet}{T}{}_{\rho}{}^{\mu\nu} + \tfrac{1}{2}\overset{\bullet}{T}{}^{\rho}{}_{\mu\nu}\overset{\bullet}{T}{}^{\nu\mu}{}_{\rho} - \overset{\bullet}{T}{}^{\rho}{}_{\mu\rho}\overset{\bullet}{T}{}^{\nu\mu}{}_{\nu}\right), \tag{C.1}$$

where $k = 8\pi G/c^4$, and

$$\overset{\bullet}{T}{}^{\rho}{}_{\mu\nu} = h_a{}^{\rho}\overset{\bullet}{T}{}^{a}{}_{\mu\nu}, \tag{C.2}$$

with

$$\overset{\bullet}{T}{}^{a}{}_{\nu\mu} = \partial_\nu h^a{}_\mu - \partial_\mu h^a{}_\nu + \overset{\bullet}{A}{}^{a}{}_{e\nu}h^e{}_\mu - \overset{\bullet}{A}{}^{a}{}_{e\mu}h^e{}_\nu \tag{C.3}$$

the torsion tensor. The gravitational field equation is obtained from the Euler-Lagrange equation for the gauge potential $B^a{}_\mu$ or, equivalently, for the tetrad field $h^a{}_\mu$:

$$\frac{\partial\overset{\bullet}{\mathscr{L}}}{\partial h^a{}_\rho} - \partial_\sigma\frac{\partial\overset{\bullet}{\mathscr{L}}}{\partial(\partial_\sigma h^a{}_\rho)} = 0. \tag{C.4}$$

The ensuing field equation can be written in the form

$$\partial_\sigma\left(h\overset{\bullet}{S}{}_a{}^{\rho\sigma}\right) - kh\overset{\bullet}{J}{}_a{}^{\rho} = 0, \tag{C.5}$$

where

$$\overset{\bullet}{S}{}_a{}^{\rho\sigma} = -\overset{\bullet}{S}{}_a{}^{\sigma\rho} \equiv -\frac{k}{h}\frac{\partial\overset{\bullet}{\mathscr{L}}}{\partial(\partial_\sigma h^a{}_\rho)} \tag{C.6}$$

is the superpotential, and

$$\overset{\bullet}{J}{}_a{}^{\rho} \equiv -\frac{1}{h}\frac{\partial\overset{\bullet}{\mathscr{L}}}{\partial h^a{}_\rho} \tag{C.7}$$

stands for the Noether gravitational energy-momentum current.

R. Aldrovandi, J.G. Pereira, *Teleparallel Gravity*, Fundamental Theories of Physics 173, DOI 10.1007/978-94-007-5143-9, © Springer Science+Business Media Dordrecht 2013

C.2 The Superpotential

Let us first obtain the explicit form of the superpotential (C.6), that is,

$$\overset{\bullet}{S}_a{}^{\rho\sigma} = -\frac{k}{h}\frac{\partial \overset{\bullet}{\mathcal{L}}}{\partial(\partial_\sigma h^a{}_\rho)}. \tag{C.8}$$

Taking the functional derivative of the first term of the lagrangian (C.1) with respect to $\partial_\sigma h^a{}_\rho$, we obtain

$$\begin{aligned}\frac{1}{4}\frac{\partial}{\partial(\partial_\sigma h^a{}_\rho)}\left(\overset{\bullet}{T}_{\lambda\mu\nu}\overset{\bullet}{T}{}^{\lambda\mu\nu}\right) &= \frac{1}{2}\overset{\bullet}{T}_b{}^{\mu\nu}\frac{\partial \overset{\bullet}{T}{}^b{}_{\mu\nu}}{\partial(\partial_\sigma h^a{}_\rho)} \\ &= \frac{1}{2}\overset{\bullet}{T}_b{}^{\mu\nu}\delta^b_a\left(\delta^\sigma_\mu\delta^\rho_\nu - \delta^\sigma_\nu\delta^\rho_\mu\right) \\ &= \overset{\bullet}{T}_a{}^{\sigma\rho}.\end{aligned} \tag{C.9}$$

In the same way, the derivative of the second term of lagrangian (C.1) yields

$$\begin{aligned}\frac{1}{2}\frac{\partial}{\partial(\partial_\sigma h^a{}_\rho)}\left(\overset{\bullet}{T}_{\rho\mu\nu}\overset{\bullet}{T}{}^{\mu\rho\nu}\right) &= \frac{1}{2}\frac{\partial \overset{\bullet}{T}{}^c{}_{\mu\nu}}{\partial(\partial_\sigma h^a{}_\rho)}\overset{\bullet}{T}{}^{\nu\mu}{}_c + \frac{1}{2}\overset{\bullet}{T}{}^c{}_{\mu\nu}\frac{\partial \overset{\bullet}{T}{}^{\nu\mu}{}_c}{\partial(\partial_\sigma h^a{}_\rho)} \\ &= \overset{\bullet}{T}{}^{\rho\sigma}{}_a - \overset{\bullet}{T}{}^{\sigma\rho}{}_a.\end{aligned} \tag{C.10}$$

Finally, the functional derivative of the last term of the lagrangian (C.1) is

$$\begin{aligned}-\frac{\partial}{\partial(\partial_\sigma h^a{}_\rho)}\left(\overset{\bullet}{T}{}^\nu{}_{\mu\nu}\overset{\bullet}{T}{}^{\lambda\mu}{}_\lambda\right) &= -2\overset{\bullet}{T}{}^{\nu\mu}{}_\nu h_b{}^\lambda\frac{\partial \overset{\bullet}{T}{}^b{}_{\mu\lambda}}{\partial(\partial_\sigma h^a{}_\rho)} \\ &= -2\overset{\bullet}{T}{}^{\nu\mu}{}_\nu h_b{}^\lambda\delta^b_a\left(\delta^\sigma_\mu\delta^\rho_\lambda - \delta^\sigma_\lambda\delta^\rho_\mu\right) \\ &= -2\overset{\bullet}{T}{}^{\nu\sigma}{}_\nu h_a{}^\rho + 2\overset{\bullet}{T}{}^{\nu\rho}{}_\nu h_a{}^\sigma.\end{aligned} \tag{C.11}$$

Combining these results, the superpotential (C.8) is found to be

$$\overset{\bullet}{S}_a{}^{\sigma\rho} = -\tfrac{1}{2}\left(\overset{\bullet}{T}_a{}^{\sigma\rho} + \overset{\bullet}{T}{}^{\rho\sigma}{}_a - \overset{\bullet}{T}{}^{\sigma\rho}{}_a - 2\overset{\bullet}{T}{}^{\nu\sigma}{}_\nu h_a{}^\rho + 2\overset{\bullet}{T}{}^{\nu\rho}{}_\nu h_a{}^\sigma\right). \tag{C.12}$$

It can also be written in the equivalent form

$$\overset{\bullet}{S}_a{}^{\sigma\rho} = \overset{\bullet}{K}{}^{\rho\sigma}{}_a + \overset{\bullet}{T}{}^{\nu\sigma}{}_\nu h_a{}^\rho - \overset{\bullet}{T}{}^{\nu\rho}{}_\nu h_a{}^\sigma, \tag{C.13}$$

with

$$\overset{\bullet}{K}{}^{\rho\sigma}{}_a = \tfrac{1}{2}\left(\overset{\bullet}{T}_a{}^{\rho\sigma} + \overset{\bullet}{T}{}^{\sigma\rho}{}_a - \overset{\bullet}{T}{}^{\rho\sigma}{}_a\right) \tag{C.14}$$

the contortion tensor. This is precisely the superpotential introduced in Eq. (9.62).

C.3 The Energy-Momentum Current

We are going now to obtain the explicit form of the gravitational energy-momentum current (C.7), that is,

$$\overset{\bullet}{j}_a{}^\rho \equiv -\frac{1}{h}\frac{\partial \overset{\bullet}{\mathcal{L}}}{\partial h^a{}_\rho}. \tag{C.15}$$

Using the lagrangian (C.1), as well as the identity

$$\frac{\partial h}{\partial h^a{}_\rho} = h h_a{}^\rho, \tag{C.16}$$

it can be rewritten in the form

$$\begin{aligned} \overset{\bullet}{j}_a{}^\rho = -\frac{1}{2k}\bigg(& \frac{1}{4}\frac{\partial \overset{\bullet}{T}{}^c{}_{\mu\nu}}{\partial h^a{}_\rho}\overset{\bullet}{T}{}_c{}^{\mu\nu} + \frac{1}{4}\overset{\bullet}{T}{}^c{}_{\mu\nu}\frac{\partial \overset{\bullet}{T}{}_c{}^{\mu\nu}}{\partial h^a{}_\rho} + \frac{1}{4}\overset{\bullet}{T}{}^c{}_{\mu\nu}\frac{\partial \overset{\bullet}{T}{}^{\nu\mu}{}_c}{\partial h^a{}_\rho} + \frac{1}{2}\frac{\partial \overset{\bullet}{T}{}^c{}_{\mu\nu}}{\partial h^a{}_\rho}\overset{\bullet}{T}{}^{\nu\mu}{}_c \\ & - \overset{\bullet}{T}{}_{\lambda\mu}{}^\lambda \frac{\partial \overset{\bullet}{T}{}^{\nu\mu}{}_\nu}{\partial h^a{}_\rho} - \frac{\partial \overset{\bullet}{T}{}_{\lambda\mu}{}^\lambda}{\partial h^a{}_\rho}\overset{\bullet}{T}{}^{\nu\mu}{}_\nu \bigg) + h_a{}^\rho \overset{\bullet}{\mathscr{L}}. \end{aligned} \tag{C.17}$$

Making then use of the torsion definition (C.3), the functional derivative that appears in the first and fourth terms of (C.17) is easily seen to be

$$\frac{\partial \overset{\bullet}{T}{}^c{}_{\mu\nu}}{\partial h^a{}_\rho} = \overset{\bullet}{A}{}^c{}_{a\mu}\delta^\rho_\nu - \overset{\bullet}{A}{}^c{}_{a\nu}\delta^\rho_\mu. \tag{C.18}$$

To compute the other functional derivatives, it will be necessary to use the properties

$$\frac{\partial h_c{}^\nu}{\partial h^a{}_\rho} = -h_a{}^\nu h_c{}^\rho \tag{C.19}$$

and

$$\frac{\partial g^{\mu\nu}}{\partial h^c{}_\rho} = -g^{\rho\nu}h_c{}^\mu - g^{\rho\mu}h_c{}^\nu, \tag{C.20}$$

which follow from the orthogonality of the tetrad and of the metric, respectively. For example, the functional derivative that appears in the second term of (C.17) can be worked out as follows:

$$\begin{aligned} \frac{\partial \overset{\bullet}{T}{}_c{}^{\mu\nu}}{\partial h^a{}_\rho} &= \eta_{cb}\frac{\partial}{\partial h^a{}_\rho}\left(g^{\mu\alpha}g^{\nu\beta}\overset{\bullet}{T}{}^b{}_{\alpha\beta}\right) \\ &= \left(\frac{\partial g^{\mu\alpha}}{\partial h^a{}_\rho}g^{\nu\beta} + g^{\mu\alpha}\frac{\partial g^{\nu\beta}}{\partial h^a{}_\rho}\right)\overset{\bullet}{T}{}_{c\alpha\beta} + \eta_{cb}g^{\mu\alpha}g^{\nu\beta}\frac{\partial \overset{\bullet}{T}{}^b{}_{\alpha\beta}}{\partial h^a{}_\rho} \\ &= \left(-g^{\nu\beta}g^{\alpha\rho}h_a{}^\mu - g^{\nu\beta}g^{\mu\rho}h_a{}^\alpha - g^{\mu\alpha}g^{\nu\rho}h_a{}^\beta - g^{\mu\alpha}g^{\beta\rho}h_a{}^\nu\right)\overset{\bullet}{T}{}_{c\alpha\beta} \\ &\quad + \eta_{cb}g^{\mu\alpha}g^{\nu\beta}\left(\overset{\bullet}{A}{}^b{}_{a\alpha}\delta^\rho{}_\beta - \overset{\bullet}{A}{}^b{}_{a\beta}\delta^\rho{}_\alpha\right) \\ &= -h_a{}^\mu\overset{\bullet}{T}{}_c{}^{\rho\nu} - g^{\mu\rho}\overset{\bullet}{T}{}_{ca}{}^\nu - g^{\nu\rho}\overset{\bullet}{T}{}_c{}^\mu{}_a - h_a{}^\nu\overset{\bullet}{T}{}_c{}^{\mu\rho} \\ &\quad + \eta_{cb}g^{\mu\alpha}g^{\nu\rho}\overset{\bullet}{A}{}^b{}_{a\alpha} - \eta_{cb}g^{\mu\rho}g^{\nu\beta}\overset{\bullet}{A}{}^b{}_{a\beta}. \end{aligned} \tag{C.21}$$

The functional derivative that appears in third term of (C.17) can be rewritten as

$$\begin{aligned} \frac{\partial \overset{\bullet}{T}{}^{\nu\mu}{}_c}{\partial h^a{}_\rho} &= \eta_{cb}\frac{\partial}{\partial h^a{}_\rho}\left(h^b{}_\sigma \overset{\bullet}{T}{}^{\nu\mu\sigma}\right) \\ &= \eta_{ca}\overset{\bullet}{T}{}^{\nu\mu\rho} + \eta_{cb}h^b{}_\sigma\frac{\partial \overset{\bullet}{T}{}^{\nu\mu\sigma}}{\partial h^a{}_\rho}. \end{aligned} \tag{C.22}$$

We have then to compute

$$\frac{\partial \dot{T}^{\nu\mu\sigma}}{\partial h^a{}_\rho} = \frac{\partial}{\partial h^a{}_\rho}\left(\eta^{bc} h_b{}^\nu \dot{T}_c{}^{\mu\sigma}\right)$$

$$= -\eta^{bc} h_a{}^\nu h_b{}^\rho \dot{T}_c{}^{\mu\sigma} + \eta^{bc} h_b{}^\nu \frac{\partial \dot{T}_c{}^{\mu\sigma}}{\partial h^a{}_\rho}$$

$$= -h_a{}^\nu \dot{T}^{\rho\mu\sigma} - \eta^{bc} h_b{}^\nu \left(h_a{}^\mu \dot{T}_c{}^{\rho\sigma} + g^{\mu\rho} \dot{T}_{ca}{}^\sigma + g^{\sigma\rho} \dot{T}_c{}^\mu{}_a + h_a{}^\sigma \dot{T}_c{}^{\mu\rho}\right)$$

$$+ \eta^{bc} h_b{}^\nu \eta_{cd} g^{\mu\alpha} g^{\sigma\rho} \dot{A}^d{}_{a\alpha} - \eta^{bc} h_b{}^\nu \eta_{cd} g^{\mu\rho} g^{\sigma\beta} \dot{A}^d{}_{a\beta}$$

$$= -h_a{}^\nu \dot{T}^{\rho\mu\sigma} - h_a{}^\mu \dot{T}^{\nu\rho\sigma} - g^{\mu\rho} \dot{T}^\nu{}_a{}^\sigma - g^{\sigma\rho} \dot{T}^{\nu\mu}{}_a - h_a{}^\sigma \dot{T}^{\nu\mu\rho}$$

$$+ h_b{}^\nu g^{\mu\alpha} g^{\sigma\rho} \dot{A}^b{}_{a\alpha} - h_b{}^\nu g^{\mu\rho} g^{\sigma\beta} \dot{A}^b{}_{a\beta}. \tag{C.23}$$

Substituting in (C.22) we get

$$\frac{\partial \dot{T}^{\nu\mu}{}_c}{\partial h^a{}_\rho} = -h_a{}^\nu \dot{T}^{\rho\mu}{}_c - h_a{}^\mu \dot{T}^{\nu\rho}{}_c - h_c{}^\rho \dot{T}^{\nu\mu}{}_a - g^{\mu\rho} \dot{T}^\nu{}_{ac}$$

$$+ h_b{}^\nu \left(g^{\mu\lambda} g^{\sigma\rho} - g^{\mu\rho} g^{\sigma\lambda}\right) \dot{A}^b{}_{a\lambda}. \tag{C.24}$$

The functional derivative appearing in the fifth term of (C.17) is

$$\frac{\partial \dot{T}^{\nu\mu}{}_\nu}{\partial h^a{}_\rho} = \frac{\partial}{\partial h^a{}_\rho}\left(g_{\nu\sigma} \dot{T}^{\nu\mu\sigma}\right)$$

$$= \frac{\partial g_{\nu\sigma}}{\partial h^a{}_\rho} \dot{T}^{\nu\mu\sigma} + g_{\nu\sigma} \frac{\partial \dot{T}^{\nu\mu\sigma}}{\partial h^a{}_\rho}.$$

Using Eq. (C.23), as well as the identity

$$\frac{\partial g_{\nu\sigma}}{\partial h^a{}_\rho} = \delta^\rho_\lambda h_{a\nu} + \delta^\rho_\nu h_{a\lambda}, \tag{C.25}$$

we get

$$\frac{\partial \dot{T}^{\nu\mu}{}_\nu}{\partial h^a{}_\rho} = -\dot{T}^{\rho\mu}{}_a - h_a{}^\mu \dot{T}^{\nu\rho}{}_\nu - g^{\mu\rho} \dot{T}^\nu{}_{a\nu} + h_b{}^\rho g^{\mu\nu} \dot{A}^b{}_{a\nu} - h_b{}^\nu g^{\mu\rho} \dot{A}^b{}_{a\nu}. \tag{C.26}$$

Finally, the functional derivative appearing in the sixth term of (C.17) is

$$\frac{\partial \dot{T}_{\lambda\mu}{}^\lambda}{\partial h^a{}_\rho} = \frac{\partial}{\partial h^a{}_\rho}\left(h_c{}^\lambda \dot{T}^c{}_{\mu\lambda}\right)$$

$$= -h_a{}^\lambda h_c{}^\rho \dot{T}^c{}_{\mu\lambda} + h_c{}^\lambda \left(\dot{A}^c{}_{a\mu} \delta^\rho{}_\lambda - \dot{A}^c{}_{a\lambda} \delta^\rho{}_\mu\right)$$

$$= -\dot{T}^\rho{}_{\mu a} + h_c{}^\rho \dot{A}^c{}_{a\mu} - h_c{}^\lambda \dot{A}^c{}_{a\lambda} \delta^\rho{}_\mu. \tag{C.27}$$

Taking into account all the results above, after some algebraic manipulations the energy-momentum current (C.15) is found to be

$$\dot{j}_a{}^\rho = \frac{1}{k} h_a{}^\lambda \dot{S}_c{}^{\nu\rho} \dot{T}^c{}_{\nu\lambda} - \frac{h_a{}^\rho}{h} \dot{\mathcal{L}} + \frac{1}{k} \dot{A}^c{}_{a\sigma} \dot{S}_c{}^{\rho\sigma}, \tag{C.28}$$

which is precisely Eq. (9.63).

Appendix D
Dirac Equation

The coupling of torsion to spinors leads to quite peculiar properties. As a support for the study of that interaction, a brief introduction to the Dirac equation is presented.

D.1 Relativistic Fields

Let us start by recalling that a field is—by definition—a relativistic field if it belongs to some representation of the Poincaré group [1], the semi-direct product of the Lorentz group $\mathscr{L}$ by the group $\mathscr{T}^{3,1}$ of translations on Minkowski spacetime:

$$\mathscr{P} = \mathscr{L} \oslash \mathscr{T}^{3,1}.$$

This is the group of isometries (or motions) on Minkowski spacetime, which means that its transformations preserve the Lorentz metric η_{ab}.

The Lorentz generators $S_{ab} = -S_{ba}$ and the translation generators P_a constitute a basis for the Lie algebra of $\mathscr{P}$, and obey the basic commutation rules

$$[p_a, p_b] = 0 \tag{D.1}$$

$$[S_{ab}, p_c] = i(\eta_{bc} p_a - \eta_{ac} p_b) \tag{D.2}$$

$$[S_{ab}, S_{cd}] = i(\eta_{bc} S_{ad} - \eta_{ac} S_{bd} - \eta_{bd} S_{ac} + \eta_{ad} S_{bc}). \tag{D.3}$$

A Lorentz transformation of the Minkowski cartesian coordinates is written as

$$x'^a = \Lambda^a{}_b x^b, \tag{D.4}$$

where $\Lambda^a{}_b$ is an element of the Lorentz group in the vector representation of generators [2]

$$(S_{ab})^c{}_d = i\big(\eta_{bd}\delta^c_a - \eta_{ad}\delta^c_b\big). \tag{D.5}$$

Under such a transformation, a general field $\Psi(x)$ will transform by the action of a representation $U(\Lambda)$,

$$\Psi'(x') = U(\Lambda)\Psi(x), \tag{D.6}$$

R. Aldrovandi, J.G. Pereira, *Teleparallel Gravity*, Fundamental Theories of Physics 173, DOI 10.1007/978-94-007-5143-9, © Springer Science+Business Media Dordrecht 2013

with

$$U(\Lambda) = \exp\left[-\frac{i}{2}\omega^{ab} S_{ab}\right]. \tag{D.7}$$

In this expression, $\omega^{ab} = -\omega^{ba}$ are the transformation parameters (3 for rotations and 3 for boosts), and S_{ab} are now the generators in the representation appropriate for the field $\Psi(x)$.

A *carrier space* is any space on which the transformations take place. The set of operators *representing* the group Lie algebra on a given carrier space is a *representation*. The Poincaré group $\mathscr{P}$ is a group of rank two. This means that there are at most two operators which commute with all its generators and are, consequently, invariant under the transformations they generate. Of course, a function of invariants is itself an invariant, so that it is possible to choose those invariant operators which have the most direct physical meaning—as will be done below. A fundamental result is that the eigenvalues of these two invariant operators classify all the representations: the members (functions, vectors, spinors, tensors, etc.) of the space carrying a representation are transformed into each other, so that those eigenvalues are kept the same under transformations.

In field theory, representations are characterized, or classified, by the eigenvalues of two carefully chosen invariant operators:

$$\eta^{ab} p_a p_b = m^2 c^2, \tag{D.8}$$

with m the mass of the particle with four-momentum p_a, and

$$\eta^{ab} W_a W_b = -m^2 c^2 s(s+1), \tag{D.9}$$

where s is the spin, and

$$W_d = -\tfrac{1}{2}\varepsilon_{abcd} S^{ab} p^c \tag{D.10}$$

is the Pauli-Lubanski operator. Each relativistic field must correspond to fixed (eigen-)values of operators $p_a p^a$ and $W_a W^a$, that is, well-defined values of mass and spin. Particles appear in field theory as the quanta of the fundamental fields. Such fields are, before quantization, wave functions $\Psi(\boldsymbol{x}, t)$ representing the states of a system in Quantum Mechanics. They are fields in the sense that $\Psi(\boldsymbol{x}, t)$ stands for a continuous infinity of possible values, one at each point of spacetime. Spacetime points $(\boldsymbol{x}, t)$ play the role of labels of the degrees of freedom Ψ.

Comment D.1 Actually, it is the covering $SL(2, C)$ of $SO(3, 1)$ which is at work, but we shall not go into these details. Let us only say that the covering is necessary to include half-integer spins. This generalizes the case of the group of usual rotations in ordinary euclidean 3-dimensional space, which is isomorphic to $SO(3)$, the group of 3×3 orthogonal matrices with determinant $+1$. This group has rank one: it has only one invariant, which is chosen to be

$$J^2 = J_x^2 + J_y^2 + J_z^2,$$

whose eigenvalues are $j(j+1)$, with j an integer. Each value of this invariant characterizes a representation. Its covering, the group $SU(2)$ of unitary 2×2 matrices with determinant $+1$, has all the representations of $SO(3)$ *plus* many others, characterized by

$$J^2 = j(j+1), \tag{D.11}$$

with j a half-integer. An example of representation of $SU(2)$, which is not a representation of $SO(3)$, is the $SU(2)$ fundamental (that is, lowest-dimensional, $j = 1/2$) representation, generated by $S_a = \sigma_a/2$, with σ_a the 2×2 Pauli matrices:

$$\sigma_1 = \begin{pmatrix} 0 & 1 \\ 1 & 0 \end{pmatrix} \quad \sigma_2 = \begin{pmatrix} 0 & -i \\ i & 0 \end{pmatrix} \quad \sigma_3 = \begin{pmatrix} 1 & 0 \\ 0 & -1 \end{pmatrix}. \tag{D.12}$$

D.2 Dirac Field

In the non-relativistic case, time evolution is ruled by the prototype of wave equation, the Schrödinger equation. It is obtained from Classical Mechanics through the so-called quantization rules, by which classical quantities become operators acting on the wave function $\psi(\boldsymbol{x}, t)$. Depending on the "representation", some quantities become differential operators and other are given by a simple product. As it is, $\psi(\boldsymbol{x}, t)$ corresponds to the configuration-space representation, in which $\boldsymbol{x}$ is the operator acting on $\psi(\boldsymbol{x}, t)$ according to the simple product rule

$$\psi(\boldsymbol{x}, t) \to \boldsymbol{x}\psi(\boldsymbol{x}, t).$$

The Hamiltonian and the three-momentum, on the other hand, are given respectively by

$$H \to i\hbar\frac{\partial}{\partial t} \quad \text{and} \quad \boldsymbol{p} \to \frac{\hbar}{i}\boldsymbol{\nabla}. \tag{D.13}$$

In the case of a free particle, for which $H = \boldsymbol{p}^2/2m$, these rules lead to the free Schrödinger equation

$$i\hbar\frac{\partial}{\partial t}\psi(\boldsymbol{x}, t) = -\frac{\hbar^2}{2m}\boldsymbol{\nabla}^2\psi(\boldsymbol{x}, t). \tag{D.14}$$

Now, the Schrödinger equation is the non-relativistic limit of the Klein-Gordon equation, which we write in the form

$$-\hbar^2\frac{\partial^2}{\partial t^2}\psi(\boldsymbol{x}, t) = -\hbar^2c^2\boldsymbol{\nabla}^2\psi(\boldsymbol{x}, t) + m^2c^4\psi(\boldsymbol{x}, t). \tag{D.15}$$

Every relativistic field must obey this equation. In fact, it corresponds to the expression

$$H^2 = \boldsymbol{p}^2c^2 + m^2c^4, \tag{D.16}$$

which is compulsory because it says simply that the field is an eigenstate of the Poincaré group invariant operator (D.8),

$$p_a p^a = \frac{H^2}{c^2} - \boldsymbol{p}^2, \tag{D.17}$$

with eigenvalue m^2c^2. Any field corresponding to a particle of mass m must satisfy this condition. Of course, once we use H^2, we shall be introducing negative energy solutions for a free system: there is no reason to exclude

$$H = -\sqrt{\boldsymbol{p}^2c^2 + m^2c^4}. \tag{D.18}$$

Following Dirac, we can look for another way to "extract the square root" of the operator H^2. In other words, we look for a linear, first-order equation both in t and in $\boldsymbol{x}$. We write [3]

$$\sqrt{\boldsymbol{p}^2c^2 + m^2c^4} = c\,\boldsymbol{\alpha}\cdot\boldsymbol{p} + \beta mc^2, \tag{D.19}$$

where $\boldsymbol{\alpha} = (\alpha_1, \alpha_2, \alpha_3)$ and β are constants to be found. Taking the square, we arrive at the conditions

$$\begin{aligned} &{\alpha_1}^2 = {\alpha_2}^2 = {\alpha_3}^2 = \beta^2 = 1 \\ &\alpha_k\beta + \beta\alpha_k = 0 \\ &\alpha_i\alpha_j + \alpha_j\alpha_i = 0 \quad (i \neq j), \end{aligned} \tag{D.20}$$

for $i, j, k = 1, 2, 3$. These conditions cannot be met if α_k and β are real or complex numbers, but can be satisfied if they are matrices. In that case, as the equation corresponding to (D.19) is the matrix equation

$$H\psi(\boldsymbol{x}, t) \equiv i\hbar\frac{\partial}{\partial t}\psi(\boldsymbol{x}, t) = \frac{\hbar}{i}c\,\boldsymbol{\alpha}\cdot\nabla\psi(\boldsymbol{x}, t) + mc^2\beta\psi(\boldsymbol{x}, t), \tag{D.21}$$

the wave function will be necessarily a column-vector, on which the matrices act. Notice that, once conditions (D.20) are satisfied, ψ will also obey the mandatory Klein-Gordon equation. We must thus look at (D.21) as an equation involving four matrices (complex, $n \times n$ for the time being) and the n-vector $\psi(x)$. As H is hermitian, so should α_k and β be:

$$\alpha_k^\dagger = \alpha_k \qquad \beta^\dagger = \beta.$$

It turns out that the minimum value of n necessary to have four matrices that are hermitian, independent and distinct from the identity is $n = 4$. Thus, α_k and β will be 4×4 matrices. The four components $\psi(x)$ correspond to particles and antiparticles with spin components $+1/2$ and $-1/2$ (whence the name "bispinor representation").

D.3 Covariant Form of the Dirac Equation

Equation (D.21) is the hamiltonian form of the Dirac equation, in which time and space play distinct roles. To go into the so-called covariant form, we first define new matrices, the celebrated Dirac's gamma matrices, as

$$\gamma^0 = \beta \quad \text{and} \quad \gamma^i = \beta\alpha^i. \tag{D.22}$$

In terms of the gamma matrices, conditions (D.20) acquire the compact form involving anti-commutators,

$$\gamma^a\gamma^b + \gamma^b\gamma^a = \{\gamma^a, \gamma^b\} = 2\eta^{ab}I, \tag{D.23}$$

where I stands for the unit 4×4 matrix. For $i = 1, 2, 3$ the matrix γ^i is anti-hermitian because

$$(\gamma^i)^\dagger = (\beta\alpha^i)^\dagger = (\alpha^i)^\dagger \beta^\dagger = \alpha^i \beta = -\beta\alpha^i = -\gamma^i, \tag{D.24}$$

whereas $\gamma^0 = \beta$ remains hermitian. Other properties are

$$(\gamma^i)^2 = -I \quad \text{and} \quad (\gamma^0)^2 = I. \tag{D.25}$$

The γ's may be taken as

$$\gamma_i = \begin{pmatrix} 0 & \sigma_i \\ -\sigma_i & 0 \end{pmatrix} \quad \text{and} \quad \gamma^0 = \begin{pmatrix} I & 0 \\ 0 & -I \end{pmatrix}, \tag{D.26}$$

in which σ_i are the Pauli matrices (D.12). This is the "Pauli-Dirac representation". Any other set of matrices γ^a obtained from these by a similarity transformation is equally acceptable [4]. We shall here only use the above representation.

The commutator of gamma matrices

$$\sigma_{ab} = \tfrac{i}{2}(\gamma_a\gamma_b - \gamma_b\gamma_a) = \tfrac{i}{2}[\gamma_a, \gamma_b] \tag{D.27}$$

has an important role. In effect, these σ_{ab} are such that

$$\left[\tfrac{1}{2}\sigma_{ab}, \tfrac{1}{2}\sigma_{cd}\right] = i\left(\eta_{bc}\tfrac{1}{2}\sigma_{ad} - \eta_{ac}\tfrac{1}{2}\sigma_{bd} + \eta_{ad}\tfrac{1}{2}\sigma_{bc} - \eta_{bd}\tfrac{1}{2}\sigma_{ac}\right). \tag{D.28}$$

Comparison with (D.3) shows that

$$S_{ab} = \tfrac{1}{2}\sigma_{ab}$$

is a Lorentz generator. Just as each matrix S_{ab} of (D.5) is a generator of the vector ($j = 1$) representation of the Lie algebra of the Lorentz group, each matrix $\sigma_{ab}/2$ is a generator of the spinor ($j = 1/2$) representation of the Lie algebra of the Lorentz group. In fact, it is found that the covariance of the Dirac equation under the transformation

$$\psi'(x') = U(\Lambda)\psi(x) \equiv \exp\left[-\tfrac{i}{4}\omega^{ab}\sigma_{ab}\right]\psi(x), \tag{D.29}$$

with $x = (\boldsymbol{x}, t)$, requires that the Dirac field $\psi(x)$ belong to this bispinor representation generated by the $\sigma_{ab}/2$. It is clear, however, that the particular form of the matrices σ_{ab} depend on the "representation" we are using for the matrices γ. In the representation (D.26) the σ_{ab} are particularly simple:

$$\sigma_{ij} = \begin{pmatrix} \varepsilon_{ijk}\sigma_k & 0 \\ 0 & \varepsilon_{ijk}\sigma_k \end{pmatrix} \quad \text{and} \quad \sigma^{0i} = \begin{pmatrix} 0 & i\sigma^i \\ i\sigma^i & 0 \end{pmatrix}. \tag{D.30}$$

Multiplying the Dirac equation (D.21) by β/c on the left, after some algebraic manipulation we find

$$i\hbar\gamma^a e_a{}^\mu \partial_\mu \psi(x) - mc\psi(x) = 0, \tag{D.31}$$

with $e_a{}^\mu$ a trivial tetrad related to the Minkowski metric:

$$\eta^{\mu\nu} = \eta^{ab} e_a{}^\mu e_b{}^\nu. \tag{D.32}$$

This is the covariant form of the Dirac equation. It follows from the lagrangian

$$\mathscr{L} = \tfrac{i}{2}\hbar c\big[\bar{\psi}\gamma^a e_a{}^\mu \partial_\mu \psi - (\partial_\mu \bar{\psi})\gamma^a e_a{}^\mu \psi\big] - mc^2 \bar{\psi}\psi, \tag{D.33}$$

where

$$\bar{\psi}(x) = \psi^\dagger(x)\gamma_0 \tag{D.34}$$

is the adjoint wave function.

Comment D.2 Due to the fact that the spinor lagrangian is linear in the field derivative, it requires the tetrad formalism. We can then say that the tetrad formalism is more fundamental than the metric one in the sense that, whereas the metric formalism can be used only in the case of tensor fields, the tetrad formalism is able to describe the field theory of both tensor and spinor fields.

D.4 Spinor Basis and Currents

Currents of Dirac fields have the general form

$$\bar{\psi}(x)\hat{\mathscr{O}}\psi(x),$$

where $\hat{\mathscr{O}}$ is an operator. Such expressions, called bilinear forms, are the only combinations that can appear in lagrangians. Only currents with a well-defined behavior under a Lorentz transformation are acceptable. From (D.29) it follows that

$$\bar{\psi}'(x')\hat{\mathscr{O}}\psi'(x') = \bar{\psi}(x)U^{-1}\hat{\mathscr{O}}U\psi(x).$$

This means that the behavior of a current is fixed by the behavior of the operator $\hat{\mathscr{O}}$ itself. It is possible to choose as a basis for the 4×4 matrices a set of matrices, *each one with a well-known behavior*. Such a set is formed by the following 16 matrices:

$$\begin{aligned}
&\Gamma^S = I\\
&\Gamma^V{}_a = \gamma_a\\
&\Gamma^T{}_{ab} = \sigma_{ab}\\
&\Gamma^P = \gamma^5\\
&\Gamma^A{}_a = \gamma^5\gamma_a,
\end{aligned}$$

where we have introduced the notation

$$\gamma^5 = \gamma_5 = i\gamma^0\gamma^1\gamma^2\gamma^3 = \begin{pmatrix} 0 & I \\ I & 0 \end{pmatrix}. \tag{D.35}$$

The superscript letter denote the corresponding behavior: scalar, vector, tensor, pseudoscalar and axial. As matrices Γ form a basis, any operator $\hat{\mathscr{O}}$ will have the form

$$\hat{\mathscr{O}} = \sum_{n=1}^{16} c_n \Gamma_n,$$

and only expressions of type

$$\bar{\psi}(x)\hat{\mathscr{O}}\psi(x) = \sum_{n=1}^{16} c_n \bar{\psi}(x)\Gamma_n \psi(x)$$

can appear in lagrangians. For example, the probability current is of the form

$$j^{\mu}(x) = \bar{\psi}(x) c \gamma^{\mu} \psi(x). \tag{D.36}$$

References

1. Bogoliubov, N.N., Shirkov, D.V.: Introduction to the Theory of Quantized Fields, 3rd edn. Wiley, New York (1980)
2. Ramond, P.: Field Theory: A Modern Primer, 2nd edn. Benjamin-Cummings, Reading (1996)
3. Bjorken, J.D., Drell, S.D.: Relativistic Quantum Mechanics. McGraw-Hill, New York (1964)
4. Bjorken, J.D., Drell, S.D.: Relativistic Quantum Fields. McGraw-Hill, New York (1965)

Index

R. Aldrovandi, J.G. Pereira, *Teleparallel Gravity*, Fundamental Theories of Physics 173,
DOI 10.1007/978-94-007-5143-9, © Springer Science+Business Media Dordrecht 2013

MIX
Papier aus verantwortungsvollen Quellen
Paper from responsible sources
FSC® C105338

If you have any concerns about our products,
you can contact us on
ProductSafety@springernature.com

In case Publisher is established outside the EU,
the EU authorized representative is:
Springer Nature Customer Service Center GmbH
Europaplatz 3, 69115 Heidelberg, Germany

Printed by Libri Plureos GmbH
in Hamburg, Germany